COMPUTER SCIENCE, TECHNOLOGY AND APPLICATIONS

MULTILAYER PERCEPTRONS

THEORY AND APPLICATIONS

COMPUTER SCIENCE, TECHNOLOGY AND APPLICATIONS

Additional books and e-books in this series can be found on Nova's website under the Series tab.

Additional in this series can be found on Nova's website under the e-book tab.

COMPUTER SCIENCE, TECHNOLOGY AND APPLICATIONS

MULTILAYER PERCEPTRONS

THEORY AND APPLICATIONS

RUTH VANG-MATA
EDITOR

Copyright © 2020 by Nova Science Publishers, Inc.

All rights reserved. No part of this book may be reproduced, stored in a retrieval system or transmitted in any form or by any means: electronic, electrostatic, magnetic, tape, mechanical photocopying, recording or otherwise without the written permission of the Publisher.

We have partnered with Copyright Clearance Center to make it easy for you to obtain permissions to reuse content from this publication. Simply navigate to this publication's page on Nova's website and locate the "Get Permission" button below the title description. This button is linked directly to the title's permission page on copyright.com. Alternatively, you can visit copyright.com and search by title, ISBN, or ISSN.

For further questions about using the service on copyright.com, please contact:
Copyright Clearance Center
Phone: +1-(978) 750-8400   Fax: +1-(978) 750-4470   E-mail: info@copyright.com.

NOTICE TO THE READER

The Publisher has taken reasonable care in the preparation of this book, but makes no expressed or implied warranty of any kind and assumes no responsibility for any errors or omissions. No liability is assumed for incidental or consequential damages in connection with or arising out of information contained in this book. The Publisher shall not be liable for any special, consequential, or exemplary damages resulting, in whole or in part, from the readers' use of, or reliance upon, this material. Any parts of this book based on government reports are so indicated and copyright is claimed for those parts to the extent applicable to compilations of such works.

Independent verification should be sought for any data, advice or recommendations contained in this book. In addition, no responsibility is assumed by the Publisher for any injury and/or damage to persons or property arising from any methods, products, instructions, ideas or otherwise contained in this publication.

This publication is designed to provide accurate and authoritative information with regard to the subject matter covered herein. It is sold with the clear understanding that the Publisher is not engaged in rendering legal or any other professional services. If legal or any other expert assistance is required, the services of a competent person should be sought. FROM A DECLARATION OF PARTICIPANTS JOINTLY ADOPTED BY A COMMITTEE OF THE AMERICAN BAR ASSOCIATION AND A COMMITTEE OF PUBLISHERS.

Additional color graphics may be available in the e-book version of this book.

Library of Congress Cataloging-in-Publication Data

ISBN: 978-1-53617-364-2

Published by Nova Science Publishers, Inc. † New York

CONTENTS

Preface		**vii**
Chapter 1	Multilayer Perceptron Artificial Neural Network: A Review *Akanksha Verma and Manoj Kumar*	**1**
Chapter 2	Machine Learning Classification for Network Centric Therapy Utilizing the Multilayer Perceptron Neural Network *Robert LeMoyne and Timothy Mastroianni*	**39**
Chapter 3	Age Estimation by Using Multi-Layer Perceptron Neural Network with Image Processing Techniques *Emre Avuçlu and Fatih Başçiftçi*	**77**
Chapter 4	Dynamic Forecasting of Electric Load Consumption Using Adaptive Multilayer Perceptron (AMLP) *Jeremias T. Lalis and Elmer A. Maravillas*	**101**

Chapter 5	Development of the Pre-Fractal Patch Antenna with Artificial Neural Network *Elder Eldervitch Carneiro de Oliveira,* *Marcelo da Silva Vieira,* *Rodrigo César Fonseca da Silva,* *Pedro Carlos de Assis Jr.* *and Paulo Fernandes da Silva Junior*	**125**
Index		**141**

PREFACE

Multilayer Perceptrons: Theory and Applications opens with a review of research on the use of the multilayer perceptron artificial neural network method for solving ordinary/partial differential equations, accompanied by critical comments.

A historical perspective on the evolution of the multilayer perceptron neural network is provided. Furthermore, the foundation for automated post-processing that is imperative for consolidating the signal data to a feature set is presented.

In one study, panoramic dental x-ray images are used to estimate age and gender. These images were subjected to image pre-processing techniques to achieve better results.

In a subsequent study, a multilayer perceptrons artificial neural network with one hidden layer and trained through the efficient resilient backpropagation algorithm is used for modeling quasi-fractal patch antennas.

Later, the authors propose a scheme with eight steps for a dynamic time series forecasting using an adaptive multilayer perceptron with minimal complexity. Two different data sets from two different countries were used in the experiments to measure the robustness and accuracy of the models.

In closing, a multilayer perceptron artificial neural network with a layer of hidden neurons is trained with the resilient backpropagation algorithm, and the network is used to model a Koch pre-fractal patch antenna.

Chapter 1 - The multilayer perceptron artificial neural network method is beneficial to solve initial value problems and boundary value problems in ordinary and partial differential equations. The artificial neural network method is an efficient method and can easily be applied to deal with the domain of higher dimensions. The approximation method based on an artificial neural network to solve the ordinary differential equation and partial differential equations are summarized, and a wide range of research works associated with these problems of differential equations of several types from the different fields are described. In this chapter, the authors present the research work done in the area of multilayer perceptron artificial neural network method for solving ordinary/partial differential equations with the authors' critical comments of study.

Chapter 2 - The application of the multilayer perceptron neural network serves an instrumental role for attaining machine learning classification accuracy in the context of Network Centric Therapy. In essence, Network Centric Therapy pertains to the use of wearable and wireless systems with the Internet of Things for the realm of biomedical engineering and healthcare. The multilayer perceptron neural network facilitates the capability of Network Centric Therapy through providing machine learning classification accuracy, which is envisioned to augment clinical situational awareness in the context of diagnosis and prognosis for patient health status in response to a therapeutic intervention. A historical perspective for the evolution of the multilayer perceptron neural network is examined. Furthermore, the foundation for automated post-processing that is imperative for consolidating the signal data to a feature set is presented. Perspectives for the application of the multilayer perceptron neural network are demonstrated for an assortment of preliminary engineering proof of concept scenarios regarding Network Centric Therapy. Future concepts for integrating the multilayer perceptron neural network with Network Centric Therapy are subsequently considered.

Preface ix

Chapter 3 - A person's identity, age or gender may need to be identified due to natural hazards such as disasters, or legal reasons such as inheritance and age manipulation. In such cases accurate information may be asked about the identity of the person from the forensic sciences. Forensic science institutions try to make the age estimation process with different human organs (example: teeth, bones etc.). These estimations are approximate estimated values. This study was conducted in order to provide the most accurate reports of forensic sciences. Forensic science makes the determination of age and gender through dental x-ray images. In this study, panoramic dental x-ray images were used to estimate age and gender. The database was created manually, with a total of 562 teeth images of 69 different dental classes. These images were first applied to image pre-processing techniques to achieve better results. After this process, the images were segmented and the feature extraction of images were made. Optional feature reduction was made. Feature vectors as a result of feature extraction process were presented as input to a multi-layered perceptron classifier. Segmentation was performed automatically and dynamically. The application was written in C # programming language. The highest 99.9% (full segment) and 100% (not full segment) classification success was achieved by using a multi-layered perceptron network.

Chapter 4 - Electric energy plays a vital role in the achievement of social economic and environment development of any nation. Thus, efficient demand planning and production of energy is needed to avoid too much over/under-estimation of electric load. In this study, the researchers proposed a scheme with eight steps for a dynamic time series forecasting using adaptive multilayer perceptron with minimal complexity. Two different data sets; each divided into three overlapping parts (training, validating and testing sets), from two different countries were used in the experiments to measure the robustness and accuracy of the models produced by the AMLP. Experiments results show the effectiveness of the proposed scheme for AMLP in forecasting the electric load consumption based on the calculated coefficient of variance of RMSD, CV (RMSD).

Chapter 5 - In this chapter, a multilayer perceptron artificial neural network with a layer of hidden neurons trained with the resilient backpropagation algorithm, *the network was used to model a Koch pre-fractal patch antenna.* The training set for the electromagnetic characterization of the antenna was obtained through simulations in the ANSYS commercial software by the momentum method. The neural network model proposed in this paper consists of a multilayer perceptron network that is able to predict antenna behavior within a region of interest with low computational cost, with a training of five thousand epoch, and means square error last than 0.0003.

In: Multilayer Perceptrons
Editor: Ruth Vang-Mata

ISBN: 978-1-53617-364-2
© 2020 Nova Science Publishers, Inc.

Chapter 1

MULTILAYER PERCEPTRON ARTIFICIAL NEURAL NETWORK: A REVIEW

Akanksha Verma and Manoj Kumar*
Department of Mathematics,
Motilal Nehru National Institute of Technology Allahabad,
Prayagraj, Uttar Pradesh, India

Abstract

The multilayer perceptron artificial neural network method is beneficial to solve initial value problems and boundary value problems in ordinary and partial differential equations. The artificial neural network method is an efficient method and can easily be applied to deal with the domain of higher dimensions. The approximation method based on an artificial neural network to solve the ordinary differential equation and partial differential equations are summarized, and a wide range of research works associated with these problems of differential equations of several types from the different fields are described. In this chapter, we present the research work done in the area of multilayer perceptron artificial neural network method for solving ordinary/partial differential equations with our critical comments of study.

Keywords: Multilayer Perceptron Artificial Neural Network, Initial Value Problem, Boundary Value Problem, Ordinary Differential Equations, Partial Differential Equations

---

*Corresponding Author's Email: manoj@mnnit.ac.in.

AMS Subject Classification: 65N

1. INTRODUCTION

An artificial neural network is a famous area of artificial intelligence research and also an abstract computational model based on the group structure of the human brain. The elementary definition of the artificial neural network is discovered by the first neurocomputers scientist," Dr. Robert Hecht-Nielsen." He defines a neural network as "A computing system made up of several simple, Highly interconnected processing elements. Which process information by their dynamic state response to external input". For more details, refer [4].

A progression of entanglements in numerous logical fields can be demonstrated with the utilization of differential equations, for example, problems in physics, chemistry, biology, economics, and engineering, etc. Because of the significance of the differential equation, we have created in the current synopsis for their solution. Main approximation methods capable of solving Differential equations are the Shooting Method, Finite DifferenceMethod (FDM), Finite Element Method (FEM), Finite Volume Approximation technique where the input data for the design of a network consists of only a set of unstructured discrete data points. In this way, the utilization of neural networks for solving differential equations can be considered as a mesh-free numerical method. This paper gives a little advance idea with respect to the development of computational investigation of the neural network, which has a ton of uses in the field of science and engineering.

The remainder of the chapter illustrates the basic fundamentals of the neural network for solving differential equations, which has been collected from many standard books related to the solution of the differential equation using neural networks and an enormous number of research articles published in reputed journals.

2. DESCRIPTION OF METHOD

In this section, we will discuss how to formulate the error function for initial/boundary value problems in ordinary differential equations. Artificial Neural Networks (ANNs) have developed as a substitute technique for the numerical solution of initial value problems (IVPs) and Boundary value problems (BVPs).

To solve the differential equations, a neural network model presents the following features:

- The solution obtained by the ANN method is differentiable and is in a closed analytic form that can be effectively utilized in any consequent calculation.

- The method can be applied in a parallel architecture.

- Solutions of differential equations obtained using neural network method has good generalization properties.

- The appropriate number of model parameters is less than any other approximation technique. In this manner, the compact solution models are obtained with very minimal required memory space.

- We can apply this method to the system characterize on either orthogonal box boundaries or irregular boundaries and realized in hardware neuroprocessor.

Further, ANN techniques produce a continuous solution over the whole domain, by removing the requirement for interpolation between fixed node points, where the solution has been gotten. The solution of IVPs and BVPs depends upon the satisfaction of given differential equation (DE) just as the fulfillment of given initial conditions (ICs) or boundary conditions (BCs). which may be a type of Dirichlet, Neumann or mixed conditions.

To solve the initial and boundary value problem, here we describe artificial neural network technique (ANNs). First, generate a trial solution of a given differential equation is composed as an addition of two sections. The first part satisfies the initial/boundary conditions and contains no flexible parameter. The second part does not contribute to the Initial/Boundary condition. The second part includes a feed-forward neural network containing a flexible parameter. We can apply this method for the solution of ordinary differential equations, the system of ordinary differential equations and partial differential equations.

In this chapter, we have given a short summary of multilayer perceptron (MLP) neural network methods for solving the differential equations. Here we have considered a multilayer artificial neural network (ANN) model with one input layer containing a single node x, that has n number of data ($x = (x_1, x_2, x_3...x_n)^T$), A hidden layer with m nodes & one output unit [4]. Figure 1 represents the structure of multilayer ANN.

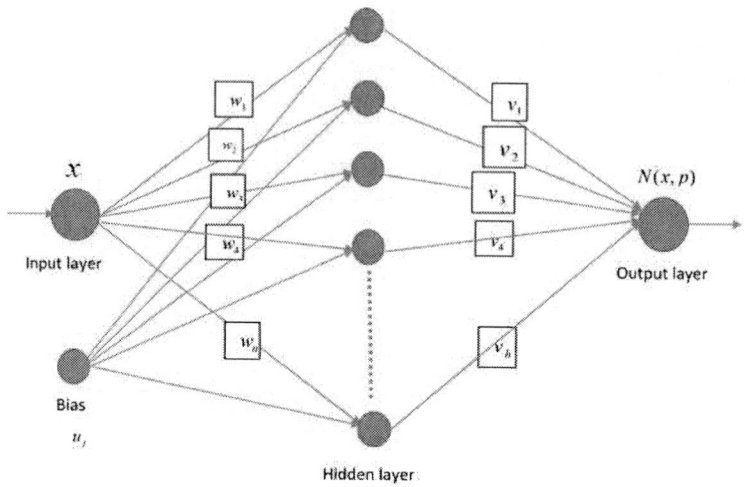

Figure 1: Structure of multilayer ANN.

An Ordinary differential equation to be solved is represented as follows: [4]

$$H(\vec{x}, \phi(\vec{x}), \nabla\phi(\vec{x}), \nabla^2\phi(\vec{x}), \ldots, \nabla^n\phi(\vec{x})) = 0, \qquad \vec{x} \in D \subseteq \mathbb{R} \quad (1)$$

subject to the definite boundary conditions. where, $D \subseteq \mathbb{R}$ shows the domain of definition, $\phi(\vec{x})$ is the solution to be calculated, H is a function that describes the framework of the differential equation, ∇ is a differential operator.

First, we discretize the domain of definition D & its boundary S into a set of distinct points $\hat{D}$ & $\hat{S}$ respectively. Then the problem of equation (1) is converted into the subsequent system of equation.

$$H(\vec{x_i}, \phi(\vec{x_i}), \nabla\phi(\vec{x_i}), \nabla^2\phi(\vec{x_i}), \ldots, \nabla^n\phi(\vec{x_i})) = 0, \qquad \forall \vec{x_i} \in \hat{D} \quad (2)$$

with constraints dictated by the boundary conditions (BCs).

Let $\phi_t(\vec{x}, \vec{p})$ is a trial solution of equation (2). where $\vec{p}$ gives the adjustable parameter like weight and biases of neural architecture. Now we transformed equation (2) into an unconstrained optimization problem as follows:

$$\min_{\vec{p}} \sum_{\vec{x_i} \in \hat{D}} (H(\vec{x_i}, \phi_t(\vec{x_i}), \nabla\phi_t(\vec{x_i}), \nabla^2\phi_t(\vec{x_i}), \ldots, \nabla^n\phi_t(\vec{x_i})))^2 \quad (3)$$

2.1. Construction of Trial Function

The trial solution $\phi_t(\vec{x}, \vec{p})$ can be constructed as follows

$$\phi_t(\vec{x}, \vec{p}) = B(x) + F(\vec{x}, N(\vec{x}, \vec{p})) \qquad (4)$$

- The trial function $\phi_t(\vec{x}, \vec{p})$ must satisfy the boundary conditions.
- It is the addition of two terms, first one is independent of flexible parameter & the other is with flexible parameter.

where, the first term $B(x)$ fulfills the initial/ boundary condition & contains no flexible parameter. The second term $F(\vec{x}, N(\vec{x}, \vec{p}))$ does not participate to the satisfaction of initial/boundary condition. $N(\vec{x}, \vec{p})$ is the single-output feed-forward neural network whose weights and biases are adjusted to minimize the error function to obtain the final ANN solution $\phi_t(\vec{x}, \vec{p})$.

To minimize the equation (3) we train the neural network. The parameters are updated during the process of training [39] & the parameter updation formula depends on the gradient of the network error with respect to network parameter.

2.2. Gradient Computation of Ordinary Differential Equation for Multilayer Perceptron

To compute the error, only network output is not sufficient but the derivatives of network output w.r.to its input is also required. The network output $N(\vec{x}, \vec{p})$ is formulated as

$$N(\vec{x}, \vec{p}) = \sum_{j=1}^{m} v_j \sigma(z_j) \qquad (5)$$

where $z_j = w_j x + u_j$
w_j is weight from input unit to hidden unit j
v_j is weight from hidden unit j to output unit
u_j is biases and $\sigma(z_j)$ is the sigmoid transfer function.

In this procedure, our aim to minimize the error function [29]. Error minimization does not only depends on the network output but also the derivative of network output with respect to input x_i and weights.

The derivative of the network output $N(\vec{x}, \vec{p})$ with respect to input x is

$$\frac{\partial N}{\partial x} = \sum_{j=1}^{m} v_j w_j \sigma^{(1)}$$

similarly the kth order derivative of N is

$$\frac{\partial^k N}{\partial x^k} = \sum_{j=1}^{m} v_j w_j^k \sigma_j^{(k)} \qquad (6)$$

let N_α is the derivative of the network output with respect to any of its input and

$$N_\alpha = D^n N = \sum_{j=1}^{m} v_j P_j \sigma_j^{\wedge} \qquad (7)$$

where

$$P_j = \prod_{k=1,2,\ldots,n} w_j^k, \qquad \wedge = \sum_{i=1}^{n} \lambda_i \qquad (8)$$

Therefore the derivatives of N_α with respect to the adjustable parameters of the network can be established as [16]

$$\frac{\partial N_\alpha}{\partial v_j} = P_j \sigma_j^{(\wedge)} \qquad (9)$$

$$\frac{\partial N_\alpha}{\partial u_j} = v_j P_j \sigma_j^{(\wedge+1)} \qquad (10)$$

$$\frac{\partial N_\alpha}{\partial w_j} = x v_j P_j \sigma_j^{(\wedge+1)} + v_j \lambda_i w_j^{(\lambda_i - 1)} \left(\prod_{k=1, k \neq i} w_j^k \right) \sigma_j^{(\wedge)} \qquad (11)$$

To minimize the error function as per the desired accuracy, now we apply Back-propagation algorithm.

2.3. Method for First Order Initial Value Problems

To demonstrate the method, we take the first order ODE [4]

$$\frac{d\phi}{dx} = g(x, \phi) \qquad \forall x \in [a, b] \qquad (12)$$

subject to initial condition $\phi(a) = A$
the ANN trial solution is written as

$$\phi_t(x, p) = A + (x - a) N(x, p) \qquad (13)$$

where $N(x,p)$ is output of feed forward network with input data $x = (x_1, x_2, ..., x_n)^T$ and parameter p. differentiating equation (13) w.r.to x

$$\frac{d\phi_t(x,p)}{dx} = (x-a)\frac{dN(x,p)}{dx} + N(x,p) \tag{14}$$

the error function for this case

$$E(x,p) = \sum_{i=1}^{m} \frac{1}{2}\left(\frac{d\phi_t(x_i,p)}{dx} - f(x_i, \phi_t(x_i,p))\right)^2 \tag{15}$$

2.4. Method for Second Order Initial Value Problems

Let us consider the general form of second order ODE

$$\frac{d^2\phi}{dx^2} = g\left(x, \phi, \frac{d\phi}{dx}\right) \qquad x \in [a,b] \tag{16}$$

subject to $\phi(a) = A$, $\phi'(a) = C$
the ANN trial solution is expressed as

$$\phi_t(x,p) = A + C(x-a) + (x-a)^2 N(x,p) \tag{17}$$

Now differentiating equation (17), we have

$$\frac{d\phi_t(x,p)}{dx} = C + 2(x-a)N(x,p) + (x-a)^2\frac{dN}{dx} \tag{18}$$

$$\frac{d^2\phi_t(x,p)}{dx^2} = 2N(x,p) + 4(x-a)\frac{dN}{dx} + (x-a)^2\frac{d^2N}{dx^2} \tag{19}$$

the error function for this case

$$E(x,p) = \sum_{i=1}^{m} \frac{1}{2}\left(\frac{d^2\phi_t(x_i,p)}{dx^2} - g\left(x_i, \phi_t(x_i,p), \frac{d\phi_t(x_i,p)}{dx}\right)\right)^2 \tag{20}$$

as discussed in above the weight from input layer to hidden layer and from hidden layer to output layer are modified according to the unsupervised back-propagation learning algorithm [29]

$$\frac{\partial E(x,p)}{\partial w_j} = \frac{\partial}{\partial w_j}\left(\sum_{i=1}^{m} \frac{1}{2}\left(\frac{d^2\phi_t(x_i,p)}{dx^2} - g\left[x_i, \phi_t(x_i,p), \frac{d\phi_t(x_i,p)}{dx}\right]\right)^2\right) \tag{21}$$

$$\frac{\partial E(x,p)}{\partial v_j} = \frac{\partial}{\partial v_j}\left(\sum_{i=1}^{m}\frac{1}{2}\left(\frac{d^2\phi_t(x_i,p)}{dx^2} - g\left[x_i, \phi_t(x_i,p), \frac{d\phi_t(x_i,p)}{dx}\right]\right)^2\right) \quad (22)$$

2.5. Method for Second Order Boundary Value Problems

Let us consider the general form of second order differential equation with boundary condition

$$\frac{d^2\phi}{dx^2} = g\left(x, \phi, \frac{d\phi}{dx}\right) \qquad x \in [a,b] \quad (23)$$

subject to boundary conditions $\phi(a) = A$ and $\phi(b) = B$
The corresponding ANN trial solution for this BVP is formulated as

$$\phi_t(x,p) = \frac{bA - aB}{b-a} + \frac{B-A}{b-a}x + (x-a)(x-b)N(x,p) \quad (24)$$

differentiating equation (24)

$$\frac{d\phi_t(x,p)}{dx} = \frac{B-A}{b-a} + (x-b)N(x,p) + (x-a)N(x,p) + (x-a)(x-b)\frac{dN}{dx} \quad (25)$$

as such, the error function is obtained as

$$E(x,p) = \sum_{i=1}^{m}\frac{1}{2}\left(\frac{d^2\phi_t(x_i,p)}{dx^2} - g\left(x_i, \phi_t(x_i,p), \frac{d\phi_t(x_i,p)}{dx}\right)\right)^2 \quad (26)$$

2.6. Method for Fourth Order Boundary Value Problems

Let us consider a fourth order differential equation

$$\frac{d^4\phi}{dx^4} = g\left(x, y, \frac{d\phi}{dx}, \frac{d^2\phi}{dx^2}, \frac{d^3\phi}{dx^3}\right) \quad (27)$$

with the following boundary condition

$$\phi(a) = \alpha, \quad \phi(b) = \beta, \quad \phi'(a) = \alpha', \quad \phi'(b) = \beta'$$

Let us assume that the ANN trial solution, which satisfy the boundary condition [4], [14]

$$\phi_t(x,p) = L(x) + M(x)N(x,p) \quad (28)$$

the trial solution satisfying the following relations

$$\begin{cases} L(a) = \alpha, \quad L(b) = \beta, \\ L'(a) = \alpha', \quad L'(b) = \beta' \\ M(a)N(a,p) = 0, \quad M(b)N(b,p) = 0, \\ M(a)N'(a,p) + M'(a)N(a,p) = 0, \\ M(b)N'(b,p) + M'(b)N(b,p) = 0 \end{cases} \quad (29)$$

the function $M(x)$ and $L(x)$ is considered as

$$M(x) = (x-a)^2(x-b)^2 Z(x) = a'x^4 + b'x^3 + c'x^2 + d'x$$

where $M(x)$ satisfy the equation (29). The motivates of introduction of $L(x)$ that is the polynomial of degree 4. where a', b', c' & d' are constants. Now from the equation (29)

$$\begin{cases} a'a^4 + b'a^3 + c'a^2 + d'a = \alpha \\ a'b^4 + b'b^3 + c'b^2 + d'b = \beta \\ 4a'a^3 + 3b'a^2 + 2c'a + d' = \alpha' \\ 4a'b^3 + 3b'b^2 + 2c'b + d' = \beta' \end{cases}$$

After solving the four equation with four unknowns. we get the general form of $L(x)$.

As such, the error function of forth order differential equation is obtained as

$$E(x,p) = \sum_{i=1}^{h} \frac{1}{2} \left(\frac{d^4 \phi_t(x_i, p)}{dx^4} - g\left[x_i, \phi_t(x_i, p), \frac{d\phi_t(x_i, p)}{dx}, \frac{d^2 \phi_t(x_i, p)}{dx^2}, \frac{d^3 \phi_t(x_i, p)}{dx^3} \right] \right)^2 \quad (30)$$

After finding the derivatives, we find the gradient of error by utilizing the error back-propagation learning technique for unsupervised training after that minimize the error function as per the desired accuracy.

3. Literature Related to Initial Value Problems and Boundary Value Problems in Ordinary Differential Equations

In this section, we described a recent development of a neural network method for solving the differential equation.

In 1990, H. Lee et al. [18] demonstrated the neural minimization algorithm for solving the differential equation by considering a finite difference equation. In 1994, A. J. Meade et al. [26, 25] have been solved linear and non-linear ordinary differential equation by developing a general numerically efficient, non-iterative feed-forward neural network method. They have been used the Hard-limit activation function, which is linear in processing time and storage, and used L_2-norm to decrease the network approximation error quadratically by multiplying the number of hidden layer neurons. They appropriated a technique from applied mathematics, perceived as a strategy for weighted residuals, and introduced how it very well may be worked to perform legitimately on the network architecture. This strategy is a summed up technique for approximating function as often as possible from given differential equations. The result acquired through the output of the network express the efficiency of approximation.

An iterative method based on the finite difference scheme for solving a class of singular boundary value problems has been presented by P.M. Lima and M. P. Carpentier in [20]. The method presented by I. E. Lagaris et al. in 1998 [16], to solve ordinary differential equations (ODEs) and partial differential equations (PDEs) with mixed (Dirichlet/Neumann) boundary condition by using the artificial neural network (ANN). To exhibit their strategy they took the accompanying general differential equation:

$$H(\vec{x}, \phi(\vec{x}), \nabla\phi(\vec{x}), \nabla^2\phi(\vec{x})) = 0, \quad \vec{x} \in D \subseteq \mathbb{R} \qquad (31)$$

subject to the specific boundary conditions.

$x = (x_1, x_2, x_3...x_n) \in \mathbb{R}^n$ shows the domain of definition and $\phi(\vec{x})$ is the solution to be calculated. The collocation method is adopted to solve the above type of differential equation (31). Authors solved this equation by constructing a trial function $\phi_t(\vec{x})$ which satisfied given boundary condition. The solution acquired is compared with the solution acquired utilizing the Galerkin finite element technique for different cases of Partial differential equations and found that this technique shows superb speculation execution.

In 2001, N. M. Duy et al. [21] showed a method to solve the linear differential equation, which depends on a multiquadric radial basis function network and categorized in the mesh-free numerical method. In 2002, L. Jianyu et al. [19] exhibited a radial basis function neural network method for solving the differential equation. Based on the thought of approximation of function or its derivatives by using a radial basis function, another new RBFN approximation procedure not quite the same as the past techniques are created. The assets of the suggested method are that it can determine all the parameters simultaneously without a learning process and it need not bother with enough data, relies on the domain and boundary. In 2003, D. R. Parisi et al. [28] introduced an amazing utilization of unsupervised neural network for solving the differential equation that implementation on a chemical engineering problem.

In 2006, A. Malek et al. [23] proposed a hybrid method dependent on the optimization technique and neural network for the solution of higher-order ordinary differential equations. They consider the general initial/boundary value problems in the form

$$\begin{cases} D\left[x, \phi, \frac{d\phi}{dx}, \frac{d^2\phi}{dx^2}, ..., \frac{d^n\phi}{dx^n}\right] = 0, & x \in [a, b] \\ C\left[x, \phi, \frac{d\phi}{dx}, \frac{d^2\phi}{dx^2}, ..., \frac{d^n\phi}{dx^n}\right] = 0, & x = a \text{ or/and } b \end{cases} \quad (32)$$

where D is the differential operator of degree n, C is initial/boundary operator. To find the approximate solution of equation(32) they consider the trial solution $\phi_t(x, p)$, where p is a flexible parameter including weight and bias in feed-forward neural network architecture and satisfy the following optimization problem:

$$\begin{cases} Min_p \int_a^b \left\| D\left[x, \phi_t(x, P), \frac{d\phi_t(x,P)}{dx}, ..., \frac{d^n\phi_t(x,P)}{dx^n}\right] \right\|^2 dx \\ C\left[x, \phi_t, \frac{d\phi_t(x,P)}{dx}, \frac{d^2\phi_t(x,P)}{dx^2}, ..., \frac{d^n\phi_t(x,P)}{dx^n}\right] = 0, & x = a \text{ or/and } b \end{cases} \quad (33)$$

The error $E(x)$ corresponding to each x must be minimized. They build up a program in MATLAB 6.5.1 to assess the conduct and qualities of the technique and conclude that the method converges to the exact and completely stable solution. In 2009, K.S. Mcfall et al. [24] has been solved with a linear differential equation on a square domain and a non-linear differential equation on a star-shaped domain with exact satisfaction by using artificial neural network method. V. Dua [7] presented the decomposition approach for parameter estimation. This methodology is established by the capacity of the artificial neural network (ANN) to approximate highly non-linear and multi input-output functions accurately. Keeping in view the applications of boundary value problems in a different branch of science and engineering, a variety of numerical methods have been discussed by various authors in the existing literature [13, 8, 22, 36, 12].

In 2005, P. Ramuhalli et al. [31] proposed a finite element neural network (FENN) method acquired by enclosing a finite element model in a neural network design that permits the quick and exact solution of the problems. To solve the forward problem, we can utilize the finite element neural network technique and in iterative algorithms, it can be utilized to find the solution of the inverse problem. The assets of this approach are that FENN can decrease the computational cost identified by utilizing the FEM in an iterative calculation for tackling inverse problems. This technique doesn't require any training, and the calculation of weight is a past procedure.

A method to choose the input variable and sparse connectivity of the lower layer of connections in feed-forward neural networks of multi-layer perceptron type is presented by H. Saxen et al. in 2006 [33]. At each step of algorithm minimum, meaningful connections are removed by a process where the lower-layer weights are zeroed, consecutively, the network corresponding to the minimum approximation error is chosen. Authors have been illustrated many problems to describe the proposed algorithm is beneficial for the consumers in obtaining appropriate input from a set of possible ones.

X. Li-Ying et al. [40] presented a new method in 2007, which is based on the cosine basis function, to solve the initial value problem in ordinary differential equations. They contrasted the outcomes and the other existing strategies and found that the created calculation is more precise than others. In 2009, J. C. Chedjou et al. [5] displayed a general idea for solving Complex/Stiff equations with the cellular neural networks (CNN) paradigm. The authors claimed that the solution is obtained faster than that with other existing methods in the literature. The differential equation is solved by I. G. Tsoulos et al. [38] in 2009 with constructed neural networks. The present method generates the trial solution in a neural network form, applying a program based on grammatical evolution. The user is just required to test the differential equations to create the test files. The final solution is showed in a closed analytic form. The proposed technique can be stretched out for the built neural systems with various transfer functions. In 2010, C. Filici [9] proposed an error estimation for the neural network method of the solution of an ordinary differential equation.

In 2013, M. Kumar et. al. [14] used the artificial neural network method for computing the buckling load of the beam-column with distinct end conditions. They modeled the problem of beam-column buckling in the differential equation mathematically. The general differential equation of beam-column is

$$EI\frac{d^4\phi}{dx^4} + F\frac{d^2\phi}{dx^2} = l \tag{34}$$

where EI represents the flexural rigidity of the beam in the plane of bending and F is an axial compressive force. Authors considered the general equation of beam-column and connected with the displacement of the center line $w(x)$ to the axial compressive force P and the lateral force $q(x)$ [32] i.e.

$$EI\frac{d^4w}{dx^4} + P\frac{d^2w}{dx^2} = q \tag{35}$$

with the boundary conditions [32, 14]

$$w(0) = w'(0) = w(k) = w'(k) = 0 \tag{36}$$

$$w(0) = w''(0) = w(k) = w''(k) = 0 \tag{37}$$

$$w(0) = w'(0) = w(k) = w''(k) = 0 \tag{38}$$

Authors have been solved a beam-column equation by using the multilayer perceptron neural network. Now from the method for fourth-order boundary value problem (27). Authors take the **Case 1** as boundary conditions given in equation (36), which means Beam column fixed at the end $x = 0$ and $x = k$. To demonstrate the method, they consider the trial solution

$$w_t(x, K) = L(x) + M(x)N(x, K) \tag{39}$$

the trial function satisfy the boundary condition given in equation (36) with the following relations

$$L(0) = 0,\ L(k) = 0,\ L'(0) = 0,\ L'(k) = 0 \tag{40}$$

$$\begin{cases} M(0)N(0, K) = 0 \\ M(k)N(k, K) = 0 \\ M(0)N'(0, K) + M'(0)N(0, K) = 0 \\ M(k)N'(k, K) + M'(k)N(k, K) = 0 \end{cases} \tag{41}$$

Where $M(x) = (x - 0)^2(x - k)^2$ and $L(x) = a'x^4 + b'x^3 + c'x^2 + d'(x)$ is a general polynomial of degree four. a', b', c' and d' are arbitrary constants. $M(x)$ satisfy the equation (40) and from the set of equations (41), we get $a' = 0$

$b' = 0$ $c' = 0$ $d' = 0$ Therefore the trial solution for beam column with both ends fixed as
$$w_t(x, K) = (x^4 + x^2 k^2 - 2x^3 k) N(x, K). \qquad (42)$$
which satisfy the boundary condition given in equation (36). The error function for the given problem of beam-column.
$$E(x_i, K) = (w_t''''(x_i, K) - g(x_i, w_t'(x_i, K), w_t''(x_i, K), w_t'''(x_i, K))) \qquad (43)$$
where,
$$w_t'(x, K) = (4x^3 + 2xk^2 - 6x^2 k)N + (x^4 + x^2 k^2 - 2x^3 k)N' \qquad (44)$$
$$w_t''(x, K) = (12x^2 + 2k^2 - 12kx)N + (8x^3 + 4xk^2 - 12x^2 k)N' + (x^4 + x^2 k^2 - 2x^3 k)N'' \qquad (45)$$

The authors take the **Case 2** as boundary conditions given in equation (37) which means is Beam column hinged at the end $x = 0$ and $x = k$. They proposed the trial solution for equation (35) with boundary condition (37)

$$w_t(x, K) = \left(\frac{16}{5} k^{-4} x^4 - \frac{32}{5} k^{-3} x^3 + \frac{16}{5} k^{-1} x \right) \left(\frac{N_0' - N_k'}{2k} x^2 - N_0' x + N \right) \qquad (46)$$

where,
$$N_0' = \left. \frac{dN}{dx} \right|_{x=0}$$
and
$$N_k' = \left. \frac{dN}{dx} \right|_{x=k}$$
trial solution satisfy the boundary condition given in (37).

$$w_t'(x, K) = \left(\frac{64}{5} k^{-4} x^3 - \frac{96}{5} k^{-3} x^2 + \frac{16}{5} k^{-1} \right) \left(\frac{N_0' - N_k'}{2k} x^2 - N_0' x + N \right) +$$
$$\left(\frac{16}{5} k^{-4} x^4 - \frac{32}{5} k^{-3} x^3 + \frac{16}{5} k^{-1} x \right) \left(\frac{N_0' - N_k'}{k} x - N_0' + N' \right) \qquad (47)$$

$$w_t''(x,K) = \left(\frac{192}{5}k^{-4}x^2 - \frac{192}{5}k^{-3}x\right)\left(\frac{N_0' - N_k'}{2k}x^2 - N_0'x + N\right)$$

$$+ \left(\frac{16}{5}k^{-4}x^4 - \frac{32}{5}k^{-3}x^3 + \frac{16}{5}k^{-1}x\right)\left(\frac{N_0' - N_k'}{k} + N''\right) \quad (48)$$

$$+2\left(\frac{64}{5}k^{-4}x^3 - \frac{96}{5}k^{-3}x^2 + \frac{16}{5}k^{-1}\right)\left(\frac{N_0' - N_k'}{k}x - N_0' + N'\right)$$

The authors take the **Case 3** as boundary condition given in equation (38) which means beam column fixed at the end $x = 0$ and hinged at the end $x = k$. They proposed the trial solution for equation (35) with boundary condition (38).

$$w_t(x,K) = sin\left(\frac{2\pi x}{k}\right)\left(\frac{N_0 - N_k}{2k}x^2 - N_0'x + xN'\right) \quad (49)$$

which satisfy the boundary condition given in equation (38).

$$w_t'(x,K) = \frac{2\pi}{k}cos\left(\frac{2\pi x}{k}\right)\left(\frac{N_0 - N_k}{2k}x^2 - N_0'x + xN'\right)$$

$$+ sin\left(\frac{2\pi x}{k}\right)\left(\frac{N_0' - N_k'}{k}x - N_0' + N'\right) \quad (50)$$

$$w_t''(x,K) = -\frac{4\pi^2}{k^2}sin\left(\frac{2\pi x}{k}\right)\left(\frac{N_0 - N_k}{2k}x^2 - N_0'x + xN'\right) + \frac{4\pi}{k}cos\left(\frac{2\pi x}{k}\right)$$

$$\left(\frac{N_0' - N_k'}{k}x - N_0' + N'\right) + \ldots sin\left(\frac{2\pi x}{k}\right)\left(\frac{N_0' - N_k'}{k} + N''\right) \quad (51)$$

In this paper [14] the authors take a Multilayer perceptron with one input layer, one hidden layer, and one output layer. To train the neural network, they used collocation points and for the minimization of error. The authors employed the back-propagation algorithm with the gradient descent method. The problem of buckling analysis of beam-column has been solved using the neural network technique by taking 10 neurons in hidden layers. The authors say that the efficiency of approximation relies upon the number of neurons in the hidden layer and does not depends on the number of hidden layers. To determine accuracy, two types of error were measured. First is absolute error and second is a relative error. The estimated value of the neural network solution has been found by the analytic solution of the equation. The authors checked the exactness of the neural network method by comparing the solution, which is hypothetically determined by Euler's technique. The completion of the above technique proves that it is quite effective as compared to other numerical methods and reduces the complexity that occurs in resolve the beam-column equation.

In 2012 M.A.Z. Raja et al. [30] have been solved one dimensional Bratu type problem by using a neural network technique dependent on inside point strategy. In 2015, M. Kumar et al. [15] presented the multilayer perceptron neural network method to find the optimal solution of Bratu type equations with less calculation effort and higher accuracy. Bratu's Problem is known as the "Liouville-Gefalnd-Bratu" problem. Bratu's problem shows up in a wide accumulation of utilization, for example, fuel start model of warm ignition, warm responses, the Chandrasekhar model of the extension of the universe. The authors take one dimensional Bratu's problem with the non-linear boundary conditions given as following:

$$\begin{cases} \phi''(x) + \alpha e^{\phi(x)} = 0 \\ \phi(0) = \phi(1) = 0, \quad 0 \leq x \leq 1 \quad and \quad \alpha > 0 \end{cases} \quad (52)$$

Equation (52) is utilized to display a combustion problem in a numerical slab. Various numerical techniques like B-spline method, Adomian decomposition method (ADM) and Finite difference method (FDM) etc. have been used to solve the Bratu's equation.

The analytic solution of the above equation is expressed as

$$\begin{cases} \phi(x) = -2 \ln\left[\dfrac{\cosh\left(\left(x-\frac{1}{2}\right)\frac{\theta}{2}\right)}{\cosh\left(\frac{\theta}{4}\right)}\right] \\ \theta = \sqrt{2\alpha}\cosh\left(\frac{\theta}{4}\right) \end{cases} \quad (53)$$

the number of solution of the Bratu type problem is depends on the critical value α_c. If $\alpha > \alpha_c$ then it has zero solution, [30] the equation has unique solution for $\alpha = \alpha_c$ and two solution when $\alpha < \alpha_c$ [35]. Where the critical value α_c satisfy the equation and the value of α_c evaluated by the following relation

$$4 = \sqrt{2\alpha_c}\, \sinh\left(\frac{\theta_c}{4}\right) \quad (54)$$

so $\alpha_c = 3.513830719$

To solve the initial/boundary value problems of linear/non-linear differential equations dependent on artificial neural networks have been widely applied [30, 15]. To introduce a reliable treatment of Bratu's problem, authors have been illustrated one initial value problem and two initial value problems of Bratu's type. They have been developed an Artificial neural network method by using log sigmoid transfer function and trained it by gradient descent algorithm to acquire closed analytic form of the solution of one dimensional first, second and initial Bratu's type problem. The computational time of the ANN method is minor than the other methods. The demonstration of the ANN method for solving the differential equation is given in references [35, 4].

For solving the equation (52) by utilizing multilayer perceptron neural system strategy, they construct a trial solution

$$\phi_t(\vec{x}, \vec{p}) = x(1-x)N(\vec{x}, \vec{p}) \qquad (55)$$

The error function for equation (52) is

$$E(\vec{x}, \vec{p}) = \sum (\phi_t''(\vec{x}, \vec{p}) - f(\phi_t'(\vec{x}, \vec{p}), \phi_t(\vec{x}, \vec{p})))^2 \qquad (56)$$

to minimize the error function, they rewrite the equation (52)

$$G = \phi_t''(\vec{x}) + \alpha e^{\phi_t(\vec{x})} \qquad (57)$$

update the parameters of neural network, they differentiate above equation (57) with respect to network parameters

$$G'(\vec{p}) = \phi_t'''(\vec{x}, \vec{p}) + \alpha e^{\phi_t(x)} \phi_t'(\vec{p}) \qquad (58)$$

In [15] the authors have been solved one initial value problem and two boundary value problem of Bratu's equation to analyse the accuracy and applicability of ANN method. They compared the exact solution of Bratu's problem with others numerical methods. Authors told the developed method is to get the number of neurons in hidden layer and training set is considered from input. They takes 11 number of neurons from the whole domain.

They first take the Bratu's initial value problem

$$\begin{cases} \phi'' - 2e^{\phi(x)} = 0 & 0 \leq x \leq 1 \\ \phi(0) = \phi'(0) = 0 \end{cases} \qquad (59)$$

trial solution for the above Bratu's problem (59)

$$\phi_t(\overrightarrow{x}, \overrightarrow{p}) = x^2 N(\overrightarrow{x}, \overrightarrow{p}) \tag{60}$$

After that, we differentiate the above trial function with respect to network parameters and minimize the error function. Authors saw that 10 neurons in hidden layer are sufficient to solve the Bratu's problem.

$$\begin{cases} \phi'' + \alpha e^{\phi(x)} = 0 & 0 \leq x \leq 1 \\ \phi(0) = 0 = \phi(1) \end{cases} \tag{61}$$

trial solution of the above Bratu's problem (61)

$$\phi_t(\overrightarrow{x}, \overrightarrow{p}) = x(1-x) N(\overrightarrow{x}, \overrightarrow{p}) \tag{62}$$

The authors applied the ANN method in equation (62) for the value of $\alpha = 1, 2$ and 3.51. Numerical simulation have been performed and result obtained using ANN scheme is comparable and more accurate than the other existing method. Authors consider the Bratu's second boundary value problem

$$\begin{cases} \phi'' + \alpha e^{-\phi(x)} = 0 & 0 \leq x \leq 1 \\ \phi(0) = 0 = \phi(1) \end{cases} \tag{63}$$

the above equation (63) is unique in relation to the standard Bratu's problem by the term $e^{-\phi(x)}$ and $\alpha > \alpha_c$ and the trial solution of Bratu's problem (64) is same as Bratu's problem (62). The second Bratu's problem is solved for the value of $\alpha = 1$ from the ANN technique.

The authors have solved these problems by using the ANN technique and variations of the Back-propagation algorithm like gradient descent technique, Levenberg-Marquardt algorithm, and conjugate gradient algorithm. They observed that the Levenberg-Marquardt algorithm given better outcomes with less registering time in comparison with the gradient descent technique.

N. Yadav et al. [41] proposed a length factor artificial neural network technique, which is possessed widely in fluid dynamics and the mass equalization of the chemical reactor. The fundamental thought process of this paper is to exhibit the artificial neural network technique for the approximate solution of the advection-dispersion equation in a steady state. By using the artificial neural network technique, they take the trial solution of advection-dispersion equation (ADE) that automatically satisfies the boundary conditions showing up in the mass equalization equation of the chemical reactor that will be advanced to fulfill the important differential equation.

A mathematical model of advection-dispersion equation for fluid flow in a chemical reactor

$$\frac{\partial c}{\partial t} = D\frac{\partial^2 c}{\partial x^2} - W\frac{\partial c}{\partial x} - \alpha c \qquad (64)$$

where,

c- Concentration (mole/m^3)
D- Dispersion coefficient (m^2/h)
W- Velocity of the water flowing (m/h)
α- First order decay coefficient (h^{-1})

If concentration (c) depend upon the space variation and does not rely upon time then

$$\frac{\partial c}{\partial t} = 0$$

and equation (64) transformed to the second order homogeneous ordinary differential equation, which represents the advection-dispersion equation in steady state

$$D\frac{\partial^2 c}{\partial x^2} - W\frac{\partial c}{\partial x} - \alpha c = 0, \quad 0 < c < L \qquad (65)$$

together with the initial or boundary conditions

$$Qc_{in} = Qc_0 - DS\frac{dc_0}{dx} \qquad (66)$$

$$\frac{dc(L,t)}{dt} = 0 \qquad (67)$$

where,

Q- flow rate (m^3/h)
S- tank's cross sectional area (m^2)
$W = \frac{Q}{S}$ (m/h) velocity of the water flowing through the tank.

To get the ANN solution of Equation (66), authors constructed trial function of the form

$$c_t(\vec{x}, \vec{p}) = A(x) + L(x)N(\vec{x}, \vec{p}) \qquad (68)$$

where,

$N(\vec{x}, \vec{p})$- ANN output
$A(x)$- Satisfy the boundary condition
L- Length factor
(1) $L = 0, \forall x$ on boundary
(2) $L \neq 0, \forall x$ within the domain
(3) $\frac{\partial L}{\partial x} \neq 0$ on the boundary

the trial solution developed in Equation (68) fulfill the boundary conditions automatically. The error function of the given advection-dispersion equation is

$$E(\vec{x}, \vec{p}) = f\left(D, W, \alpha, x, c_t, \frac{dc_t}{dx}, \frac{d^2 c_t}{dx^2}\right) \quad (69)$$

to minimize the error function $E(\vec{x}, \vec{p})$ and iteratively update the ANN weights $\vec{p}$, they have been used gradient descent algorithm. For detailed description of ANN technique refer [41]. The ANN element contains a group of ANNs that are enhanced with the distinct training points n, the number of neurons in hidden layer H and weights are taken arbitrarily. For the suitability of the method authors take a reactor which length is $L = 10m$ with a chemical injected at the steady-state at a constant concentration of $c_{in} = 100 mol/m^3$ and the speed of water through the tank $W = 1m/h$, $\alpha = 0.2 h^{-1}$ and dispersion coefficient $D = 1, 4, 0.1 m^2/h$.

In order to study the behavior of the advection-dispersion equation (ADE), they consider three cases with the different value of dispersion coefficients and For the training of the neural system authors take the following training points $n = 10, 20, 50, 100$ and the number of neurons in hidden layer is in equidistant from $H = 5, 10, 15, 20, 25, 30, 35$ with 25 different set of random starting weights.

For the primary case ($D = 1m^2/h$), the normal behavior of fluid flow in a chemical reactor, the concentration diminishes to break up the response as the fluid in the cylinder. In the case, where a chemical of large dispersion coefficient is introduced $D = 4(m^2/h)$, the concentration curve tends to straighten, and the rate of change of concentration reduces along the longitudinal axis of the reactor. when the dispersion coefficient is enough reduced i.e., $D = 0.1(m^2/h)$, the concentration curve becomes much steeper as mixing is less less significant with respect to shift in weather conditions. To check the exactness of the ANN technique authors contrasted the ANN solution with the exact solution just as the Finite difference method (FDM) solution. They found that the ANN solution shows a nearby concurrence with the exact solution inside the the domain.

In 2016, H.F. Parapari et al. [27] presented a novel framework for solving a nonlinear ordinary differential equation by using a neural network. They solved the nonlinear differential equations by neural network technique and compared it to the analytic solution. The assets of this technique are fast convergence speed and useful to all of the differential equations. In this article, the authors introduced the structure of feed-forward neural networks and describe the uses of neural networks is modeling function and solving differential equations. Werbos and Rumelhart developed a back-propagation algorithm to trained the neural network. This algorithm separated into two sections: propagation and weight update.

They consider the general form of differential equation with initial/boundary condition as follows:

$$\begin{cases} D\left[x, \phi, \frac{d\phi}{dx}, \frac{d^2\phi}{dx^2}, ..., \frac{d^n\phi}{dx^n}\right] = 0, & x \in [a, b] \\ C\left[x, \phi, \frac{d\phi}{dx}, \frac{d^2\phi}{dx^2}, ..., \frac{d^n\phi}{dx^n}\right] = 0, & x = a \text{ or}/\text{and } b \end{cases} \quad (70)$$

Let the approximate solution of problem (70) in form of $\phi_t(x, p)$, where p is the flexible parameter. An approximate solution y_t to optimize the weight bias is unknown values.

$$\underline{E} = \underline{e} = -\alpha \nabla_e F(\underline{e}) \quad (71)$$

$$lim_{k \to \infty} \underline{e}[k] = 0 \quad (72)$$

the complication of finding the numerical solution of equation (70) on points at regular interval is similar to find the function that fulfills the conditions for the unconstrained optimization problem.

$$W(k+1) = W(k) - \alpha_1 sgn(\underline{e})\underline{p}$$
$$\underline{b}(k+1) = \underline{b}(k) - \alpha_2 sgn(\underline{e})\underline{p} \quad (73)$$

To demonstrate the method they take first and second order ODE as following:

$$\frac{d\phi(x)}{dx} = g(x, \phi) \quad x \in [a, b] \quad (74)$$

with initial condition $\phi(a) = A$.

To take care of this issue, first, we break the interval $[a, b]$ in N points and the placement of the points in the vector $\underline{p}$ or $\underline{x}$, where the initial conditions are the matrix element of weight and bias. The approximate error is given as

$$\underline{e} = \frac{d\phi}{dx} - g(x, \phi) \tag{75}$$

To improve the $\frac{d\phi}{dx}$, adjust the weight and bias of the neural network. By computing the errors, parameters of the neural system have proceeded to the next iterations. After a suitable number of iteration, the network will reach the final solution for solving ODE's so that the norm of the error vector tends to zero. A similar procedure for solving second-order ordinary differential equation they have been described.

In this article [27] the authors conclude this technique has been presented for solving the nonlinear ODE with the back-propagation algorithm by using a multilayer perceptron. They provide two examples and show that this technique has high convergence speed and high accuracy as compared to the analytic solution of the ODE.

In 2015, S. Ezadi et al. [3] introduced a neural network method based on the Semi-Taylor Series for solving an ordinary differential equation. In this approach author take the trial solution of the given differential equation, which is the addition of two-part, the first part contains the Semi-Taylor series with no adjustable parameter and the other part includes the neural network and adjustable parameter (weight, bias). The new approach gives the solution with high accuracy in comparison to the neural network technique.

They take a new model as initial value first order ODE

$$\begin{cases} \frac{d\phi(x)}{dx} = f(x, \phi(x)), & x \in [a, b] \\ \phi(a) = A \end{cases} \tag{76}$$

The trial solution of the problem (76)

$$\phi_t(x, p) = U(x) + xN(x, p) \tag{77}$$

The Taylor series expansion of $\psi(x)$, which is analytic in a neighbor-hood of real and complex number is the power series

$$\phi(x) = \phi(x_0) + \frac{\phi'(x_0)(x - x_0)}{1!} + \frac{\phi''(x_0)(x - x_0)^2}{2!} + \frac{\phi'''(x_0)(x - x_0)^3}{3!} + \ldots \tag{78}$$

They have been written it in form of sigma

$$\phi(x) = \sum_{n=0}^{\infty} \frac{\phi^{(n)}(x_0)}{n!}(x - x_0)^n$$

and the exponential function

$$e^x = \sum_{n=0}^{\infty} \frac{\phi^{(n)}}{n!}$$

Praise ordinary function on Basis Semi-Taylor series

$$\phi(x) = \phi(x_0) + \frac{\phi(x_0)(x - x_0)}{1!} + \frac{\phi(x_0)(x - x_0)^2}{2!} + \frac{\phi(x_0)(x - x_0)^3}{3!} + \ldots$$

Now praise order function on basis Semi-Taylor

$$U(x) = \left(1 + \frac{x}{1!} + \frac{x^2}{2!} + \frac{x^3}{3!}\right)\phi(x_0) \tag{79}$$

put the value of (79) in trial solution (77)

$$\phi_t(x, p) = \left(1 + \frac{x}{1!} + \frac{x^2}{2!} + \frac{x^3}{3!}\right)\phi(x_0) + xN(x, p) \tag{80}$$

Differentiating (80) w.r.to x

$$\phi'_t(x, p) = \left(1 + x + \frac{x^2}{2}\right)\phi(x_0) + N(x, p) + x\frac{\partial N}{\partial x} \tag{81}$$

The author takes a multilayer perceptron neural network, which have one hidden layer with H sigmoid units and a linear output unit. where

$$N = \sum_{i=1}^{H} v_i \sigma(z_i), \quad z_i = w_i(x+1)\epsilon + u_i, \quad \epsilon > 0. \tag{82}$$

In this article [27] authors have been used linear and sigmoid transfer function. The output of the sigmoid function belongs to the interval [0 1], since this is differentiable function.

The sigmoid function

$$\sigma(z_i) = \frac{1}{1 + e^{-z_i}}$$

First derivative of sigmoid function

$$\sigma'(z_i) = -\sigma^2 + \sigma$$

Differentiating N w.r.to input x_i, is

$$\frac{\partial N}{\partial x} = \sum_{i=1}^{H} v_i w_i \left[\frac{1}{1+e^{-w_i(x+1)\epsilon + b_i}} - \left(\frac{1}{1+e^{-w_i(x+1)\epsilon + b_i}} \right)^2 \right] \qquad (83)$$

Corresponding error function

$$E = \sum_{i=1}^{m} \left(\phi'_t(x_i, p) - g(x_i, \phi_t(x_i, p)) \right)^2 \qquad (84)$$

They have been used quasi-Newton BFGS (Broyden Fletcher Goldfarb Shanno) method for minimizing the error function.

To show the performance and characteristics of this new technique, they discuss one example and simulation is conducted on MATLAB 12, and the initial weights are randomly selected. The author outlines the first run through the capacity of the neural system and the Semi-Taylor series to accurate the solutions of an ordinary differential equation. The main purpose of using the Semi-Taylor Series with Neural networks is that the applicability in function approximation.

4. Literature Related to Initial Value Problems and Boundary Value Problems in Partial Differential Equations

In 1994, M. W. M. G. Dissanayake et al. [6] solved a linear Poisson equation and thermal conduction with a non-linear heat equation. In 1998, I. E. Lagaris et al. [16] presented a neural network technique for solving an ordinary differential equation as well as a partial differential equation. In this article, they have discussed only two dimensional PDE problems with Dirichlet boundary conditions and mixed boundary conditions (with Neumann). They have been discussed the Poisson equation.

$$\frac{\partial^2 \phi(x,t)}{\partial x^2} + \frac{\partial^2 \phi(x,t)}{\partial t^2} = f(x,t) \qquad (85)$$

$x \in [0,1]$ $\phi \in [0,1]$ with Dirichlet boundary condition
$\phi(0,t) = f_0(t)$, $\phi(1,t) = f_1(t)$ $\phi(x,0) = g_0(x)$ $\phi(x,1) = g_1(x)$
The trial solution of Poisson equation

$$\phi_T(x,t) = A(x,t) + x(1-x)t(1-t)N(x,t,\vec{P}) \qquad (86)$$

where $A(x,t)$ is selected as it satisfy the boundary condition

$$A(x,t) = (1-x)f_0(t) + xf_1t + (1-t)\{g_0(x) - [(1-x)g_0(0) \\ + xg_0(1)]\} + t\{g_1(x) - [(1-x)g_1(0) + xg_1(1)]\} \qquad (87)$$

For mixed boundary condition of the form

$$\phi(0,t) = f_0(t) \quad \phi(1,t) = f_1(t) \quad \phi(x,0) = g_0(x) \text{ and } \frac{\partial \phi(x,1)}{\partial t} = g_1(x)$$

where, Dirichlet conditions on the part of boundary and Neumann conditions elsewhere.

The trial solution with respect to mixed boundary conditions

$$\phi_T(x,t) = C(x,t) + x(1-x)t\left[N(x,t,\vec{P}) - N(x,1,\vec{P}) - \frac{\partial N(x,1,\vec{P})}{\partial t}\right] \qquad (88)$$

where,

$$C(x,t) = (1-x)f_0(t) + xf_1(t) + g_0(x) - [(1-x)g_0(0) \\ + xg_0(1)] + t\{g_1(x) - [(1-x)g_1(0) + xg_1(1)]\} \qquad (89)$$

The error function of the above PDE is

$$E(\vec{P}) = \sum_i \left\{\frac{\partial^2 \phi(x_i,t_i)}{\partial x^2} + \frac{\partial^2 \phi(x_i,t_i)}{\partial t^2} - f(x_i,t_i)\right\}^2 \qquad (90)$$

where x_i, t_i are points in $[0,1] \times [0,1]$.

In this paper [16] they have been given three examples to clarify the ANN method gives a higher accurate solution as compared to the finite element method (FEM) and all problems have been defined on the domain $[0,1] \times [0,1]$. They take a multilayer perceptron with two input, ten sigmoid hidden units, and one linear output unit. Author's told that the neural network technique gives solutions of better interpolation accuracy and consider a mesh of 10×10 points while in the finite element method 18×18 mesh has employed. The quantity of parameters standardized in x-pivot, and in the neural methodology, the genuine number of parameters is $20x$, while in the finite-element techniques are $225x$. From the above assumptions, it ought to be noticed that the neural strategy is brilliant and the exactness of this technique can be dealt with by expanding the neurons in the hidden layer. In the neural system, strategy time increments straightly with the quantity of parameters, while in the FEM case, the time scale quadratically. For solving the neural algorithm authors have been utilized Sun Ultra Sparc workstation with 512Mb of primary memory.

To solve time dependent partial differential equation with initial/boundary condition by using neural networks Authors presented an optimization technique in [34]. They have been used minimization technique and collocation method to find the approximate solution in closed analytic form of time dependent PDE with boundary conditions of the following form:

$$\begin{cases} \forall i_1 = 1, ..., p_1 : D_{i_1}\left[t, x, ..., \frac{\partial^{\alpha_0+\alpha_1+...+\alpha_n}}{\partial t^{\alpha_0}\partial x_1^{\alpha_1}...\partial x_n^{\alpha_n}} y_i(x,t), ...\right] = 0, \; t \in [t_0, t_{max}], x \in \Omega \\ \forall i_2 = 1, ..., p_2 : I_{i_2}\left[t_0, x, ..., \frac{\partial^{\alpha_0+\alpha_1+...+\alpha_n}}{\partial t^{\alpha_0}\partial x_1^{\alpha_1}...\partial x_n^{\alpha_n}} y_i(x,t_0), ...\right] = 0, \; x \in \Omega \; 1 \leq i \leq m \\ \forall i_3 = 1, ..., p_3 : B_{i_3}\left[t, x, ..., \frac{\partial^{\alpha_0+\alpha_1+...+\alpha_n}}{\partial t^{\alpha_0}\partial x_1^{\alpha_1}...\partial x_n^{\alpha_n}} y_i(x,t), ...\right] = 0, \; t \in [t_0, t_{max}], x \in \partial\Omega \end{cases}$$
(91)

where D_{i_1}, I_{i_2} and B_{i_3} are real valued multi-variable functions and system of partial differential equations are non-linear time-dependent initial boundary conditions respectively. x is real valued spatial variable, t is the time. $\Omega \subseteq \mathbb{R}^n$ is a bounded domain, $\alpha_0, \alpha_1, ..., \alpha_n \in \mathbb{N}_0^{n+1}$ ($\mathbb{N}_0 = NU0$) is a multi-index variable and $\phi(t,x) = [\phi_1(t,x), ..., \phi_m(t,x)]$ is the unknown real vector-valued function over the cylindrical domain $[t_0, t_{max}] \times \Omega \subseteq \mathbb{R}^n$.

As stated by Kolmogorov and Cybenko theorems, which include 3 layered feed-forward neural systems with customizable parameters for the solution. In this article, writers comprehended the 2-D biharmonic equation in a rectangular domain with Neumann/Dirichlet boundary conditions by constructing a specific trial solution. In the minimization process, they used the Nelder-Mead simplex method, which does not involve gradient. The ANN solution is in closed analytic form and the assets of this technique are that we can apply it to solve the non-linear time-dependent system of PDEs and the technique is summed up for tackling the higher-order and non-linear issue.

I. E. Lagaris et al. in 2000, [17] have been solved boundary value problems with the irregularly shaped region. They studied Partial differential equations with the case of complicated boundary conditions, where a series of consecutive points dictate the boundary. They take the PDEs of the form

$$L\phi = f \qquad (92)$$

For solving these type of equations collocation method is approved, and then the problem is changed over into an unconstrained optimization problem. A model dependent on the collaboration of two feed-forward neural systems of two distinct sorts are created, which can be composed as,

$$\phi_M(\vec{x},p) = N(\vec{x},p) + \sum_{l=1}^{M} q l e^{-\lambda|\vec{x}-\alpha\vec{r}l+\vec{h}|^2} \qquad (93)$$

The addition in the above equation shows a radial basis function network with M hidden nodes and all offer a typical exponential factor λ. For minimizing the error function they utilized the combination of the gradient method and penalty method. which is beneficial and results are quite encouraging. They tried their strategy on these types of problems and get amazing outcomes.

In 2001, L.P. Aarts et al. [1] presented a neural network method to solve a partial differential equation. They have been taken more than one input with a single hidden layer, single-output without bias are efficient for approximating the function and its derivatives. To explain the methodology, they take the initial value problem

$$\frac{d^2\phi}{dx^2} + \phi = 0, \quad t \in [0,1]$$
$$\frac{d\phi}{dt} = 1, \quad \phi(0) = 0 \qquad (94)$$

After that, They obtained a specifically structured network. To obtain the solution of PDE and its initial/boundary condition they use an efficient algorithm and trained all the network simultaneously and explain their method by taking two examples. They observe that a more accurate solution can be obtained if take the value of parameters in the interval [-5, 5]. Consequences of applying this technique for one and two dimensions problem are excellent in the sense of exactness, convergence, stability, and proficiency.

To solve a dimension quasi-linear PDE Burger equation, the Authors have been presented a modified neural network method in [10]. This technique is direct relevance to the nth order PDE. They examined the Burger equation of the form

$$z_t + zz_x = yz_{xx} \quad a < x < b, \ t > 0 \tag{95}$$

with the accompanying initial and boundary condition

$z(x, 0) = G(x) \quad a < x < b$
$z(a, t) = g_1(t), \ z(b, t) = g_2(t), \ t > 0$

This is one-dimension quasi-linear parabolic PDE and $y > 0$ is the coefficient of kinematic viscosity of the fluid. In this modified approach, the neural network is trained in an unsupervised manner utilizing an error function that is derived from the differential equation itself and the governing boundary conditions. They trained the network by gradient descent back-propagation method with a non-negative error function. In this article [10] it can be concluded that the ANN solutions are viewed as good and increasingly exact as compared to the solution obtained by other numerical techniques.

N. Sukavanam et al. [37] in 2003, a controlled heat problem up to 3 decimal points was solved utilizing 3 layered and feed-forward neural networks. They exhibit a computational technique to solve boundary control of a semi-linear controlled heat equation and designed controller based on an artificial neural network. The neural network approximates control terms and a trial solution is written for the controlled heat equation as a sum of two terms. They examined the controlled semi-linear heat equation in Q (with control applied on the boundary) is

$$\begin{aligned} \frac{\partial u}{\partial t} - c^2 \Delta u &= \phi(u) \quad in \ Q \\ u(0) &= u_0 \quad in \ \Omega \\ u &= g \quad in \ \Sigma \end{aligned} \tag{96}$$

where Ω is a bounded domain in $\mathbb{R}^n$ with the smooth boundary Γ. Let the time interval $(0, T)$ and $Q = \Omega \times (0, T)$, $\Sigma = \Gamma \times (0, T)$. $u = u(x, t)$ denotes the temperature at $x \in \Omega$ and time $t \in [0, T]$, ϕ is a non-linear operator satisfying Lipschitz continuity, Δ denotes the Laplace operator, c is real constant and g represents boundary control function.

They describe an approach for computing boundary control of linear as well as semi-linear heat equation based on ANN. First, they discretize the domain Q and boundary Σ and then constructed a trial solution of a given heat equation. Boundary controls are given in terms of the feed-forward neural network, after that, they transform the problem of the system of the equation to an unconstrained optimization problem. To train the network, they used the Back-propagation algorithm and the sigmoid activation function. To explain the procedure, they have been solved one- dimensional controlled non-linear heat equation with control applied on the boundary and Two- dimensional controlled linear heat equation in $\mathbb{R}^2$ on the unit square. The solution acquired by the given methodology demonstrates that the strategy is viable and given accuracy of order 10^{-3}. This strategy can be applied to tackling both linear and non-linear state equations.

In 2010 M. Baymani et al. [2] developed a new method for solving the Stokes problem, which is based on neural network technique. To get the solution of the Stokes problem, the author breaks the given problem into 3 independent Poisson equations and after solved these three problems, we get the solution of Stoke's problem. For checking the accuracy of the results, the author compared the obtained solution with the exact solution and other existing methods. They saw that the ongoing new system has higher exactness as compared to other and the number of model parameters required is less than other approximation methods.

Stokes equations depict the movement of a liquid in R^n (n=2 or 3). These equations are to be solved for an obscure velocity vector $u(x,y) = (u_i(x,y))_{1 \leq i \leq n} \in R^n$ and pressure $p(x,y) \in R$

$$\begin{cases} \Delta u_1 + \frac{\partial p}{\partial x} = f_1 & in \ \Omega \subset R^n \\ -\Delta u_2 + \frac{\partial p}{\partial y} = f_2 & in \ \Omega \\ \frac{\partial u_1}{\partial x} + \frac{\partial u_2}{\partial y} = 0 & in \ \Omega \end{cases} \tag{97}$$

with boundary conditions:
$$u = (u_1, u_2) = (u_1^0, u_2^0) = u^0, \text{ on } \partial\Omega$$

To apply ANN technique and solve the problem (97) authors have been applied the operators $\frac{\partial}{\partial x}$ and $\frac{\partial}{\partial y}$ on the first and second equations individually and get:

$$-\Delta \left(\frac{\partial u_1}{\partial x} + \frac{\partial u_2}{\partial y} \right) + \frac{\partial^2 p}{\partial x^2} + \frac{\partial^2 p}{\partial y^2} = (f_1)_x + (f_2)_y \tag{98}$$

Using third equation of (97) in (98)

$$\frac{\partial^2 p}{\partial x^2} + \frac{\partial^2 p}{\partial y^2} = (f_1)_x + (f_2)_y \quad (99)$$

Now the Stokes problem transform in Poisson equation and it has infinitely many solution. By introducing boundary conditions, they have been found the approximate solution of (99) by ANN technique.

The trial solution of Poisson equation can be written as

$$p_t(x, y) = A(x, y) + x(1-x)y(1-y)N(x, y, P) \quad (100)$$

where $A(x)$ is developed as to fulfill the boundary conditions,

$$A(x, y) = (1-x)h_0(y) + xh_1(y) + (1-y)g_0(x) - [(1-x)g_0(0) + xg_0(1)] \\ + yg_1(x) - [(1-x)g_1(0) + xg_1(1)] \quad (101)$$

where $h_0(y) = p(0, y)$, $h_1(y) = p(1, y)$, $g_0(x) = p(x, 0)$ & $g_1(x) = p(x, 1)$ the second term of the trial solution doesn't influence the boundary condition, It evaporates at the piece of the boundary where Dirichlet boundary conditions are introduced. The error function of the above PDE problem

$$E(p) = \sum_i \left\{ \frac{\partial^2 p_t(x_i, y_i)}{\partial x^2} + \frac{\partial^2 p_t(x_i, y_i)}{\partial y^2} - F(x_i, y_i) \right\}^2 \quad (102)$$

where $F = (f_1)_x + (f_2)_y$ and (x_i, y_i) is a point in Ω.

By solving unconstrained optimization problem (102), the weight parameters are obtained and then the trial solution for p_t is obtained. By replacing the value of p_t in first equation of problem (97), we get

$$\frac{\partial^2 u_1}{\partial x^2} + \frac{\partial^2 u_1}{\partial y^2} = \frac{\partial(p_t)}{\partial x} - f_1 \quad (103)$$

This is the Poisson equation for u_1. By replacing p_t from u_{1t} and F from $\frac{\partial(p_t)}{\partial x} - f_1$, in equation (102), they get the optimization problem for u_{1t}. Proceeding in similar way they found the optimization problem for u_{2t}. Author presented one example to demonstrate the method, they used one input layer with two input node, one hidden layer with 5 hidden node and one linear output node.

For a given input vector $x = (x_1, x_2)$, the output of network is

$$N(x,p) = \sum_{i=1}^{5} v_i \sigma(z_i)$$

where $z_i = \sum_{j=1}^{2} u_i + w_{ij} x_j$ and $\sigma(z)$ is sigmoid activation function.

The exact analytic solution is already known, so they test the accuracy of the obtained solution by taking the value of u_1, u_2 & p and choose the value of f_1, f_2.

After obtaining the solution of the Stokes problem, the authors compare it with Aman-Kerayechian method for 25 points and see that the ANN technique has a little error. They told the assets of ANN method, By ANN method they find the solution of the Stokes problem on every point of training interval and after solving the problem, found an approximate function for the solution by which they can ascertain the appropriate response at each point effectively.

In 2015, E. A. Hussain et al. [11] presented a new approach based on modified artificial neural network and optimization technique to solve the partial differential equation. According to the new approach, the training point ought to be chosen over an open interval without training the network in the first and last element of the interval. Therefore the calculation is a less and computational error is reduced. In this article, the introduced method is demonstrated by two numerical examples. The authors proposed a modified artificial neural network method for solving a partial differential equation that relies upon the capacity estimation effectiveness of the feed-forward neural system and results in the improvement of a solution written in closed analytic form. The proposed method to modify the ANN & based on replaced every x in the input vector $\vec{x} = (x_1, x_2, \ldots, x_n)$, $x_j \in [a, b]$ by one single degree polynomial.

$$Q(x) = \epsilon(x+1), \quad \epsilon \in (0,1) \tag{104}$$

So the new input vector will be: $(Q(x_1), Q(x_2), \ldots, Q(x_n))$, $Q(x_j) \in (a, b)$, These training points chose over an open interval (a, b) without training the neural network in first and last point. So, the figuring volume including computational error is reduced.

For a given input vector $(x_1, x_2, \ldots, x_n)$, $x_j \in [a, b]$, the output of this modified artificial neural network

$$N = \sum_{i=1}^{H} v_i \sigma(z_i) \tag{105}$$

where $z_i = \sum_{j=1}^{n} w_{ij}Q(x_j) + b_i$, $\sigma(z_i)$ is hyperbolic tangent activation function, and $Q(x_j) = \epsilon(x_j + 1)$, $\epsilon \in (0,1)$, & $x_j \in [a,b]$ this implies $Q(x_j) \in (a,b)$

Differentiating N w.r. to weight parameters as

$$\frac{\partial N}{\partial v_i} = \sigma(w_{ij}Q(x_j) + b_i) = \sigma(\epsilon(x_j+1)w_{ij} + b_i) \qquad (106)$$

$$\frac{\partial N}{\partial b_i} = v_i\sigma'(w_{ij}Q(x_j) + b_i) = v_i\sigma'(\epsilon(x_j+1)w_{ij} + b_i) \qquad (107)$$

$$\frac{\partial N}{\partial w_{ij}} = v_iQ(x_j)\sigma'(w_{ij}Q(x_j) + b_i) = v_iQ(x_j)\sigma'(\epsilon(x_j+1)w_{ij} + b_i) \qquad (108)$$

The authors have solved partial differential equations with Dirichlet or Neumann boundary conditions. All problems defined in the domain $[0,1] \times [0,1]$ and for training divided into 10 equidistant point in $[0,1]$. For each entry x and y, the input neurons roll out no improvements in its input. so the input to the hidden neurons is

$$Net_j = xw_{j1} + yw_{j2} + B_j, \quad j = 1, 2, ...m \qquad (109)$$

where w_{j1} & w_{j2} are weights from input layer to j^{th} unit of hidden layer. B_j is an jth bias for the jth unit in hidden layer. The output in the Hidden neuron is:

$$z_j = \sigma(net_j), \quad j = 1, 2, ...m \qquad (110)$$

For modified artificial neural network

$$net_j = Q(x)w_{j1} + Q(y)w_{j2} + B_j = \epsilon(x+1)w_{j1} + \epsilon(y+1)w_{j2} + B_j \qquad (111)$$

$$z_j = \sigma(Q(x)w_{j1} + Q(y)w_{j2} + B_j) \qquad (112)$$

$$N = \sum_{j=1}^{m} v_j\sigma(\epsilon(x+1)w_{j1} + \epsilon(y+1)w_{j2} + B_j) \qquad (113)$$

where $j = 1, 2, ...m, \epsilon \in (1,0)$ and $Q(x), Q(y) \in (a,b)$

In this paper [11] the authors have been solved two problems. The initial one is Laplace equation with Dirichlet boundary condition and the other one is Non-linear Partial differential equation with mixed boundary condition. They report some Numerical results & in all cases utilized three-layer feed-forward neural system with two information units, one hidden layer with 10 hidden units and one output unit, and hyperbolic tangent activation function, $\sigma(x) = \frac{e^x - e^{-x}}{e^x + e^{-x}}$, For each problem the analytic solution is known, so test the accuracy of the obtained solution by computing the deviation $\Delta u(\vec{x}) = |u_t(\vec{x}) - u_a(\vec{x})|$. For minimizing the error function they used BFGS quasi-Newton method. The numerical solution found by the modified artificial neural network has been compared with the usual artificial neural network. So the authors presented a hybrid approach based on a modified artificial neural network for solving the Partial differential equation. The logic for utilizing a modified artificial neural network is their applicability in function approximation.

ACKNOWLEDGMENTS

The authors are grateful to the National Board of Higher Mathematics (NBHM), Government of India for providing financial support to carry out this work through its project sanctioned Order No. 2/48(1)2016 R&D II/6824.

CONCLUSION

This paper concerned with the current review of multi-layer perceptron (MLP) neural network method and its advantage for solving linear and non-linear ordinary and partial differential equation with initial/boundary condition. The approximation method based on artificial neural network for solving ordinary differential equations & partial differential equation is summarized and a large amount of research paper associated with this problem of differential equations of several types from different fields was analyzed. We accept that a lot of work has been done in various fields. When we partner an artificial neural system to this marvel, it adds many interesting highlights to the solution that is the acquired ANN solution displays a smooth approximation that can be classified and differentiated continuously on the domain. A significant property of the ANN technique is that it doesn't expect linearization to take care of a non-linear problem.

REFERENCES

[1] Lucie P. Aarts and Peter Van Der Veer. Neural network method for solving partial differential equations. *Neural Processing Letters*, 2001.

[2] Modjtaba Baymani, Asghar Kerayechian, and Sohrab Effati. Artificial Neural Networks Approach for Solving Stokes Problem. *Applied Mathematics*, 2010.

[3] Fen Bilimleri Dergisi, Ezadi Somayeh, Sahar Askari, and Mitra Jasemi. Cumhuriyet Üniversitesi Fen Fakültesi Numerical Solution of ordinary Differential Equations Based on Semi-Taylor by Neural Network improvement. *Cumhuriyet University Faculty of Science Science Journal (CSJ)*, 36(3):36, 2015.

[4] Snehashish Chakraverty and Susmita Mall. *Artificial neural networks for engineers and scientists: Solving ordinary differential equations*. 2017.

[5] J.C. Chedjou, K. Kyamakya, M.A. Latif, U.A. Khan, I. Moussa, and Do Trong Tuan. *Solving stiff ordinary differential equations and partial differential equations using analog computing based on cellular neural networks*. 2015.

[6] M. W.M.G. Dissanayake and N. Phan Thien. Neural - network - based approximations for solving partial differential equations. *Communications in Numerical Methods in Engineering*, 1994.

[7] Vivek Dua. An Artificial Neural Network approximation based decomposition approach for parameter estimation of system of ordinary differential equations. *Computers and Chemical Engineering*, 2011.

[8] Vivek Dua and Pinky Dua. A simultaneous approach for parameter estimation of a system of ordinary differential equations, using artificial neural network approximation. *Industrial and Engineering Chemistry Research*, 2012.

[9] Cristian Filici. Error estimation in the neural network solution of ordinary differential equations. *Neural Networks*, 2010.

[10] Hayati, Mohsen and Behnam Karami. Feedforward NN for Solving PDE. *Journal of Applied Sciences*, 7(19):2812–2817, 2007.

[11] Eman Hussian and Mazin Suhhiem. Numerical Solution of Fuzzy Partial Differential Equations by Using Modified Fuzzy Neural Networks. *British Journal of Mathematics & Computer Science*, 12(2):1–20, 2015.

[12] Manoj Kumar and Garima Mishra. An Introduction to Numerical Methods for the Solutions of Partial Differential Equations. *Applied Mathematics*, 02(11):1327–1338, 2011.

[13] Manoj Kumar and Neha Yadav. Multilayer perceptrons and radial basis function neural network methods for the solution of differential equations: A survey. *Computers and Mathematics with Applications*, 2011.

[14] Manoj Kumar and Neha Yadav. Buckling analysis of a beam-column using multilayer perceptron neural network technique. *Journal of the Franklin Institute*, 2013.

[15] Manoj Kumar and Neha Yadav. Numerical Solution of Bratu's Problem Using Multilayer Perceptron Neural Network Method. *National Academy Science Letters*, 2015.

[16] Isaac Elias Lagaris, Aristidis Likas, and Dimitrios I. Fotiadis. Artificial neural networks for solving ordinary and partial differential equations. *IEEE Transactions on Neural Networks*, 1998.

[17] Isaac Elias Lagaris, Aristidis C. Likas, and Dimitrios G. Papageorgiou. Neural-network methods for boundary value problems with irregular boundaries. *IEEE Transactions on Neural Networks*, 2000.

[18] Hyuk Lee and In Seok Kang. Neural algorithm for solving differential equations. *Journal of Computational Physics*, 1990.

[19] Qi Yingjian Li Jianyu, Luo Siwei and Huang Yaping. Numerical solution of differential equations by radial basis function neural networks. pages 1–12, 2002.

[20] P. M. Lima and M. P. Carpentier. Iterative methods for a singular boundary-value problem. *Journal of Computational and Applied Mathematics*, 111(1-2):173–186, 1999.

[21] N Mai-Duy and T Tran-Cong. Numerical solution of differential equations using multiquadric radial basis functions networks. *Neural networks : the official journal of the International Neural Network Society*, 2001.

[22] Ms Sonali B Maind and Ms. Priyanka Wankar. Research Paper on Basic of Artificial Neural Network. *International Journal on Recent and Innovation Trends in Computing and Communication*, 2014.

[23] A. Malek and R. Shekari Beidokhti. Numerical solution for high order differential equations using a hybrid neural network-Optimization method. *Applied Mathematics and Computation*, 2006.

[24] Kevin S. McFall and James Robert Mahan. Artificial neural network method for solution of boundary value problems with exact satisfaction of arbitrary boundary conditions. *IEEE Transactions on Neural Networks*, 2009.

[25] A. J. Meade and A. A. Fernandez. The numerical solution of linear ordinary differential equations by feedforward neural networks. *Mathematical and Computer Modelling*, 1994.

[26] Andrew J Meade and Alvaro A Fernandez. Solution of Nonlinear Ordinary Differential Equations By Feedforward Neural Networks Solution of Nonlinear Ordinary Differential Equations By Feedforward Neural Networks. 20(713):19–44, 1892.

[27] Hamed Fathalizadeh Parapari and Mohammad Bagher Menhaj. Solving nonlinear ordinary differential equations using neural networks. *2016 4th International Conference on Control, Instrumentation, and Automation, ICCIA 2016*, (January):351–355, 2016.

[28] Daniel R. Parisi, María C. Mariani, and Miguel A. Laborde. Solving differential equations with unsupervised neural networks. *Chemical Engineering and Processing: Process Intensification*, 2003.

[29] Sehgal Er. Parveen, Gupta Sangeeta, and Kumar Dharminder. Minimization of Error in Training a Neural Network Using Gradient Descent Method. *International Journal of Technical Research*, 1(1):10–12, 2012.

[30] Muhammad Asif Zahoor Raja and Siraj ul Islam Ahmad. Numerical treatment for solving one-dimensional Bratu problem using neural networks. *Neural Computing and Applications*, 2014.

[31] Pradeep Ramuhalli, Lalita Udpa, and Satish S. Udpa. Finite-element neural networks for solving differential equations. *IEEE Transactions on Neural Networks*, 2005.

[32] Goutam Saha and Sajeda Banu. Buckling Load of a Beam-Column for Different End Conditions Using Multi-Segment Integration Technique. *ARPN Journal of Engineering and Applied Sciences*, 2007.

[33] H. Saxén and F. Pettersson. Method for the selection of inputs and structure of feedforward neural networks. *Computers and Chemical Engineering*, 2006.

[34] R. Shekari Beidokhti and A. Malek. Solving initial-boundary value problems for systems of partial differential equations using neural networks and optimization techniques. *Journal of the Franklin Institute*, 2009.

[35] Yazdan Shirvany, Mohsen Hayati, and Rostam Moradian. Numerical solution of the nonlinear Schrodinger equation by feedforward neural networks. *Communications in Nonlinear Science and Numerical Simulation*, 2008.

[36] Randhir Singh, Abdul Majid Wazwaz, and Jitendra Kumar. An efficient semi-numerical technique for solving nonlinear singular boundary value problems arising in various physical models. *International Journal of Computer Mathematics*, 2016.

[37] N. Sukavanam and Vikas Panwar. Computation of boundary control of controlled heat equation using artificial neural networks. *International Communications in Heat and Mass Transfer*, 2003.

[38] Ioannis G. Tsoulos, Dimitris Gavrilis, and Euripidis Glavas. Solving differential equations with constructed neural networks. *Neurocomputing*, 2009.

[39] L. Wang and J.M. Mendel. *Structured trainable networks for matrix algebra*. 2002.

[40] Li Ying Xu, Hui Wen, and Zhe Zhao Zeng. The algorithm of neural networks on the initial value problems in ordinary differential equations. *ICIEA 2007: 2007 Second IEEE Conference on Industrial Electronics and Applications*, pages 813–816, 2007.

[41] Neha Yadav, Kevin Stanley McFall, Manoj Kumar, and Joong Hoon Kim. A length factor artificial neural network method for the numerical solution of the advection dispersion equation characterizing the mass balance of fluid flow in a chemical reactor. *Neural Computing and Applications*, 2018.

In: Multilayer Perceptrons
Editor: Ruth Vang-Mata

ISBN: 978-1-53617-364-2
© 2020 Nova Science Publishers, Inc.

Chapter 2

MACHINE LEARNING CLASSIFICATION FOR NETWORK CENTRIC THERAPY UTILIZING THE MULTILAYER PERCEPTRON NEURAL NETWORK

Robert LeMoyne, PhD and Timothy Mastroianni
Department of Biological Sciences, Northern Arizona University,
Flagstaff, Arizona, US;
Cognition Engineering, Pittsburgh, Pennsylvania, US

ABSTRACT

The application of the multilayer perceptron neural network serves an instrumental role for attaining machine learning classification accuracy in the context of Network Centric Therapy. In essence, Network Centric Therapy pertains to the use of wearable and wireless systems with the Internet of Things for the realm of biomedical engineering and healthcare. The multilayer perceptron neural network facilitates the capability of Network Centric Therapy through providing machine learning classification accuracy, which is envisioned to augment clinical situational awareness in the context of diagnosis and prognosis for patient health status in response to a therapeutic intervention. A historical perspective for the evolution of the multilayer perceptron neural network is examined. Furthermore, the foundation for automated post-processing that is imperative for consolidating the signal data to a feature set is presented. Perspectives for the application of the multilayer perceptron neural network are demonstrated for an assortment of preliminary engineering proof of concept scenarios regarding Network Centric Therapy. Future concepts for integrating the multilayer perceptron neural network with Network Centric Therapy are subsequently considered.

Keywords: multilayer perceptron neural network, machine learning, network centric therapy, cloud computing environment, wearable inertial sensors, wireless inertial sensors, accelerometer, gyroscope, patellar tendon reflex, reflex response, gait, movement disorder, Parkinson's disease, Essential tremor, deep brain stimulation

INTRODUCTION

Machine learning, such as through the multilayer perceptron neural network, is central to the concept of Network Centric Therapy. Another inherent aspect of Network Centric Therapy is the wearable and wireless inertial sensor system. The combination of these attributes elucidate the eminent and rampant presence of the Internet of Things for the biomedical community and associated healthcare [1, 2, 3, 4, 5, 6].

The progressive evolution of the inertial sensor systems that are wearable and capable of wireless access has transitioned to accessibility with Cloud computing resources [1, 2, 3, 4, 5, 6]. Preliminary demonstration involved the application of locally wireless inertial sensor systems with connectivity to proximally situated personal computers and inertial sensors requiring manual data transfer to personal computers [7, 8]. The following transitional evolution was the application of smartphones and portable media devices for with effective Cloud computing access through access to email [9, 10, 11, 12, 13, 14, 15].

Recently the capability for wearable and wireless systems to access a Cloud computing environment has been demonstrated with segmented wireless connectivity to a smartphone and tablet [16, 17]. The wearable and wireless inertial sensor systems enable the collection of objectively quantified data to characterize human movement [1, 2, 4, 5, 7, 8, 9, 10, 11, 12, 13, 14, 15]. Another transitional evolution has been the adaptation of machine learning classification, such as the multilayer perceptron neural network, from descriptive and inferential statistics to distinguish between various health status scenarios [18, 19, 20].

The amalgamation of these foundations for Network Centric Therapy has lead to the development of diagnostic capability through the machine learning classification of inertial sensor signal data, which is consolidated to a feature set through software automation. Research, development, testing, and evaluation has successfully demonstrated the efficacy of the synergy of Network Centric Therapy with the multilayer perceptron neural network as a preferred machine learning algorithm. The scope of these endeavors pertains to the differentiation of a hemiplegic affected and unaffected limb pair for scenarios, such as gait and reflex response [1, 2, 18]. Network Centric Therapy with the multilayer perceptron neural network has been also successfully conducted for distinguishing between deep brain stimulation for the treatment of movement disorders, such as Parkinson's disease and Essential tremor [4, 5, 19].

These Network Centric Therapy machine learning classification endeavors utilize the multilayer perceptron neural network through the Waikato Environment for Knowledge Analysis (WEKA) [18, 19]. A general perspective of the process for applying machine learning classification though WEKA is provided, such as the collection of inertial sensor signal data through wearable and wireless systems, the consolidation of the signal data to a feature set, and machine learning classification through WEKA.

MULTILAYER PERCEPTRON NEURAL NETWORK

The brain demonstrates the capacity to perceive between distinct scenarios. The fundamental basis for the capability to attain classification accuracy between distinguishable scenarios is derived from the neuronal level. This observation from neurology has been a foundation for the development of machine learning classification algorithms, such as the multilayer perceptron neural network, which is a computational semblance of the neuron [21].

A further consideration of the neuron is warranted, in order to better understand the multilayer perceptron neural network. The neuron is composed of three primary components: dendrites, soma, and axon. The dendrite is the input source for neural signals, which are conveyed and integrated at the soma. The soma is the body of the neuron. Subsequently, the processed neural signal is conveyed through the axon for transmission to a source, such as another dendritic process [22, 23].

The Waikato's Environment for Knowledge Analysis (WEKA) features a multitude of machine learning classification algorithms, for which the multilayer perceptron neural network is predominant. The multilayer perceptron neural network provides a computation representation of the neuron as it consists of three layers (input, hidden, and output). Intrinsic to WEKA the multilayer perceptron neural network applies a quantity of input layer nodes that is equivalent to the number of feature set attributes. The output layer nodes equals the number of classes to be distinguished between, for which as classification accuracy is derived. For the multilayer perceptron neural network of WEKA the number of hidden nodes is equivalent to the sum of the total number of input layer nodes and total number of output layer nodes with the sum then divided by two. A representative multilayer perceptron neural network is presented in Figure 1 [24, 25, 26].

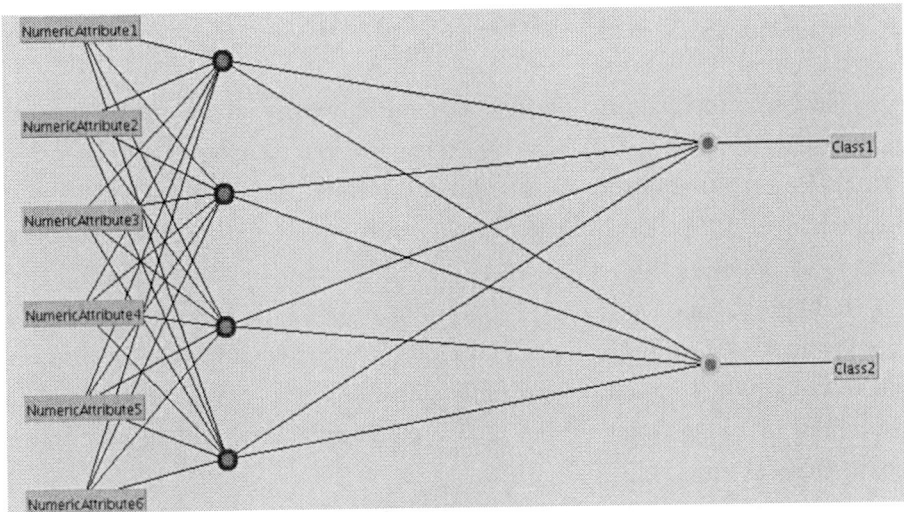

Figure 1. A generic multilayer perceptron neural network consisting of an input layer (feature set numeric attributes), hidden layer ([input layer nodes + output layer nodes]/2), and output layer (classes) [24, 25, 26].

The input layer receives the quantified numeric attributes through a feature set. The hidden layer receives weighted input derived from the input layer to establish the output layer. The output layer represents the distinct classes prescribed in the feature set [21, 27].

From a neurological perspective the input layer emulates the dendritic aspect of the neuron. The hidden layer represents the soma of the neuron. The output layer can interpreted as the axonal process emanating an output for the hidden layer.

The multilayer perceptron neural network algorithm utilized by WEKA acquires a neural network that achieves an optimal classification accuracy. With the general structure of the multilayer perceptron neural network defined as a relation for the input layer nodes and output layer nodes to derive the number of hidden layer nodes established, the proper determination of weights for the hidden layer needs to be ascertained. The algorithm for defining the weights for the hidden layer within the neural network is called backpropagation [27].

BACKPROPAGATION

The backpropagation algorithm was preliminarily advocated by Bryson and Ho in 1969 [28, 29]. The backpropagation technique applies error which is backpropagated in the direction from the output layer to the hidden layer [28]. This concept was incorporated into neural networks during the 1980's [30].

Backpropagation utilizes an algorithm referred to as gradient descent that is used for optimization. Inherent to this technique is the use of derivatives, which precludes a simple step function. Another function that is similar to a step function, but is also differentiable would achieve these mathematical criteria [27].

$$f(x) = \frac{1}{1 + e^{-x}}$$

Figure 2. The sigmoid function equation [27].

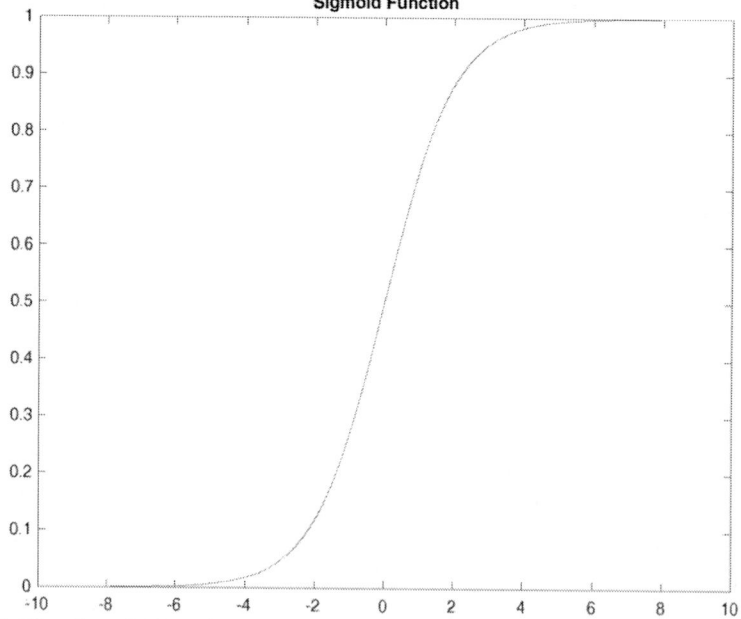

Figure 3. The sigmoid function graphical representation [27].

A sigmoid function satisfies these criteria of resembling the shape of a step function while being differentiable. For large negative inputs the sigmoid function converges to zero. For large positive inputs the sigmoid function approaches one. The equation of the sigmoid function is presented in Figure 2. A graphical representation of the sigmoid function is displayed in Figure 3 [27].

FURTHER MATTERS FOR THE MULTILAYER PERCEPTRON NEURAL NETWORK

The multilayer perceptron neural network is a well known form of the neural network, which is a feed-forward network. For this technique the output is based on the current input instance, which is not considered cyclical. Other methods, such as the recurrent neural network, illustrate cyclical properties. This involves feedback from the neural network that is influenced by computational contributions derived from previous input. This feature equips the recurrent neural network with the impression of memory [27].

The multilayer perceptron neural network also has some inherent disadvantages. The conventional gradient descent algorithm for the multilayer perceptron neural network protracts the convergence of the process for achieving optimal classification accuracy. Another concern is the means of attaining the finalized machine learning model is obscured [27].

These concerns for the multilayer perceptron neural network can impact its suitability for various configurations of Network Centric Therapy. There are two configuration strategies for Network Centric Therapy. The first configuration would apply transmission of wearable and wireless system data to the Cloud computing environment with machine learning classification attained by the Cloud. The second configuration would involve machine learning classification at the wearable and wireless system level with data conveyed to a Cloud computing environment [4, 31, 32].

The first configuration would be more appropriate for the application of the multilayer perceptron neural network, because the Cloud computing environment would be equipped with considerable processing capability. This processing capability would ameliorate computational requirements to converge the machine learning algorithm. The secondary configuration would be more appropriate for machine learning classification algorithms that require less computationally intensive. The first configuration would include the time to transmit data to the Cloud computing environment and receive machine learning classification status from the Cloud computing environment, but this observation is considered appropriate for Network Centric Therapy [4, 31, 32].

The other concern pertains to the matter of the clinician trusting the results of the machine learning model with respect to the opacity of the algorithm process. As machine learning becomes more integrated within the biomedical community this subject is envisioned to progressively ameliorate. The use of machine learning, such as through the multilayer perceptron neural network, is envisioned to substantially advance clinical acuity for diagnosis, therapeutic intervention, and prognosis of patient health. The multilayer perceptron neural network has been successfully applied from the preliminary perspective of Network Centric Therapy to achieve considerable classification accuracy for differentiating various health status scenarios [1, 2, 4, 5, 9, 18, 19].

Preliminary Application of Network Centric Therapy through the Multilayer Perceptron Neural Network using the Smartphone

The smartphone has been successfully demonstrated as a wearable and wireless inertial sensor system using machine learning classification, such as the multilayer perceptron neural network. This established preliminary proof of concept for Network Centric Therapy. For example, the experimentation site and post-processing site have been situated on opposite sides of the continental United States [1, 2, 4, 5, 9, 14, 15, 18, 19].

EVOLUTIONARY PATHWAY OF THE MULTILAYER PERCEPTRON NEURAL NETWORK FOR DIFFERENTIATING HEMIPLEGIC REFLEX PAIRS THROUGH WEARABLE AND WIRELESS SYSTEMS

The origins of Network Centric Therapy derive from the evolution of the wireless quantified reflex system. Traditionally an ordinal scale has been applied to quantify tendon reflex response, but this strategy's reliability has been a subject of contention [33, 34, 35, 36, 37]. Previous attempts have been made to quantify the tendon reflex response [33, 38, 39, 40, 41, 42, 43, 44, 45, 46, 47, 48, 49]. The wireless quantified reflex system developed over the course of four incremental evolutionary improvements that eventually lead to the ability to quantify patellar tendon reflex response and latency in an accurate, reliable, and reproducible manner. The key features of the wireless quantified reflex system are an impact pendulum and a wireless accelerometer system that was effectively wearable [33, 50, 51, 52, 53, 54, 55, 56, 57].

Eventually the wireless quantified reflex system was applied to identify quantified disparity for a hemiplegic patellar tendon reflex pair [58, 59]. Further transitional evolution applied the smartphone and portable media device in the context of wearable and wireless inertial sensor systems. This development utilized a unique aspect of Network Centric Therapy, which is the ability to geographically separate the experimental location from the post-processing resources [60, 61, 62, 63].

The combination of the amalgamation of smartphones and portable media devices with the perceptible quantified disparity of hemiplegic patellar tendon reflex pairs established the basis for actively pursuing machine learning. Machine learning algorithms were envisioned to establish classification accuracy to distinguish between the affected leg and unaffected leg in terms of reflex response. The first application with machine learning incorporated the wireless reflex quantification system using a portable media device as a wireless accelerometer system to distinguish between a hemiplegic affected patellar tendon reflex and unaffected patellar tendon reflex. This endeavor incorporated the support vector machine using WEKA [64].

Further improvements of this application were pending. The multilayer perceptron neural network was later applied because of its resemblance to the foundation of neurological perceptivity [21]. Furthermore, the gyroscope signal intrinsic to the smartphone and portable media device was considered to be a more clinically pertinent signal [65, 66].

The wireless quantified reflex system equipped with a smartphone as a functional wireless gyroscope platform acquired quantified reflex response for a hemiplegic affected leg and unaffected leg with respect to the patellar tendon reflex. The gyroscope signal was post-processed through software automation to consolidate a feature set based on the following numeric attributes:

- Maximum gyroscope signal
- Minimum gyroscope signal
- Time differential between maximum and minimum

With these attributes establishing the feature set the multilayer perceptron neural network achieved considerable classification accuracy to distinguish between the hemiplegic affected and unaffected reflex pair [67].

A similar themed research endeavor applied supramaximal stimulation of the patellar tendon reflex through a manually operated reflex hammer. A portable media device was incorporated as a functional wireless gyroscope platform. Figure 4 presents the WEKA Explorer for the machine learning process. Figure 5 demonstrates the resultant multilayer perceptron neural network that attained considerable classification accuracy to distinguish between the patellar tendon reflex responses of the hemiplegic reflex pair [68].

Machine Learning Classification for Network Centric Therapy ... 49

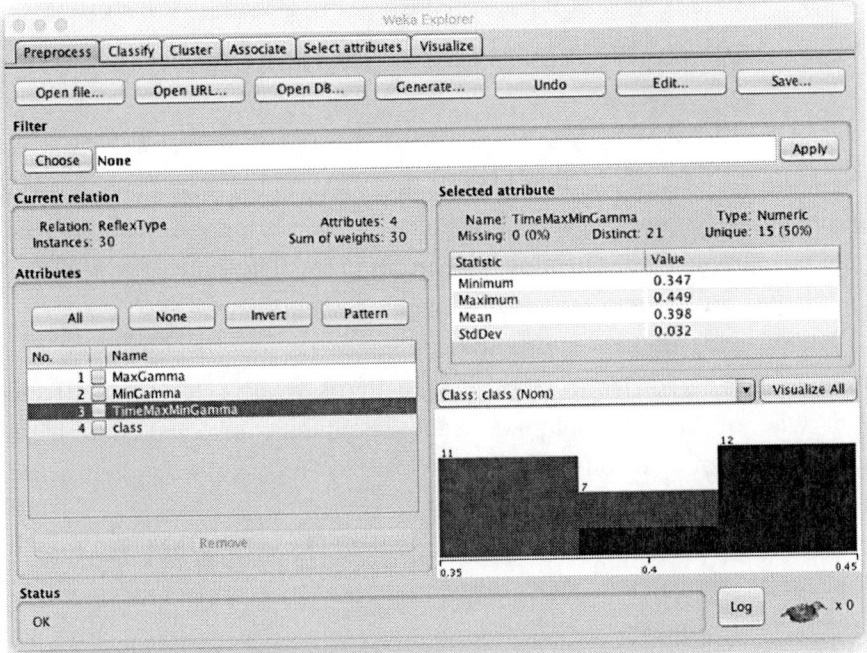

Figure 4. The WEKA Explorer for deriving machine learning classification accuracy to differentiate between a hemiplegic patellar tendon reflex pair [68].

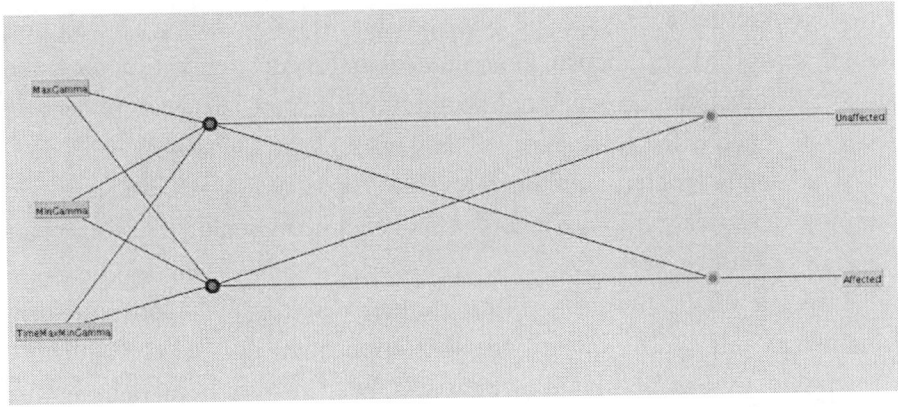

Figure 5. The multilayer perceptron neural network for establishing classification accuracy to distinguish between a hemiplegic patellar tendon reflex pair [68].

MULTILAYER PERCEPTRON NEURAL NETWORK FOR DIFFERENTIATING HEMIPLEGIC GAIT

The tendon reflex serves a unique role for the synchronous activity of gait [33, 50, 69, 70]. Intuitively, the success of machine learning classification for distinguishing between reflex response for hemiparesis could be extrapolated to hemiplegic gait. In particular, topics of interest are hemiplegic gait and reduced arm swing due to hemiparesis.

Hemiparesis imparts substantial influence on gait. The affected limb produces notably different movement patterns relative to the unaffected limb. One of the contributors to the hemiplegic affected side is the manifestation of spasticity [69, 70]. Wearable and wireless systems, such as the smartphone and portable media device, have been successfully applied for quantifying gait [71, 72, 73, 74, 75, 76, 77, 78, 79]. Beyond the scope of descriptive and inferential statistics, machine learning classification serves as a logical extrapolation for a more advanced technique to distinguish between the unaffected leg and hemiplegic affected leg during gait [80].

The combination of wearable and wireless systems with functional Cloud computing access can be augmented with the multilayer perceptron neural network. In essence an experimental gait evaluation site, such as the subject's home, can be separated geographically from the post-processing location, which is the foundation of Network Centric Therapy [1, 2, 3, 4, 5, 6]. The amalgamation demonstrates the capacity to achieve machine learning classification to distinguish between a hemiplegic lower limb pair during gait [80].

Using a smartphone as a wearable and wireless gyroscope platform hemiplegic gait can be quantified. This application utilizes the clinically perceptible gyroscope signal, which can be acquired by mounting the smartphone about the lateral malleolus of the ankle joint by an elastic band. Because the experiment applies a single smartphone, gait speed should be maintained as constant for both respective legs of the hemiplegic pair. In order to maintain constant gait speed, a treadmill is implemented [80].

With software automation a feature set is consolidated from the gyroscope signal data. The feature set is comprised of descriptive statistics to assemble the feature set numeric attributes. Using the multilayer perceptron neural network from WEKA considerable classification accuracy is attained to distinguish between a hemiplegic affected lower limb and the unaffected lower limb [80].

Another manifestation of hemiplegic gait asymmetry is known as reduced arm swing. During gait the arms naturally swing, but for hemiparesis the affected arm displays a notable impairment of arm swing compared to the unaffected arm [81]. This scenario can also be quantified using the smartphone as a wearable and wireless gyroscope platform, which constitutes a preliminary representation of Network Centric Therapy [1, 2, 3, 4, 5, 6].

The smartphone can be secured about the wrist of each arm for acquiring the gyroscope signal that quantifies the characteristics of arm swing. With the signal data transmitted wirelessly to the Internet as an email attachment, post-processing can be conducted anywhere in the world with Internet access. Software automation consolidates the gyroscope signal data to a feature set consisting of numeric attributes based on descriptive statistics. Using the multilayer perceptron neural network considerable classification accuracy is attained [81].

Another condition that demonstrates reduced arm swing is Erb's palsy, which involves impairment to the peripheral nerves of the upper arm. As an alternative the portable media device is applied as a wearable and wireless gyroscope platform with essentially the same software platform as the smartphone. With access to a local wireless Internet zone the data is conveyed wirelessly to the Internet through an email attachment for pending post-processing [82].

The feature set is comprised of numerical attributes represented by descriptive statistics. Figure 6 presents the WEKA Explorer for the machine learning endeavor. Figure 7 demonstrates the resultant multilayer perceptron neural network that attained considerable classification accuracy to differentiate reduced arm swing for a subject with an arm affected by Erb's palsy relative to an unaffected arm [82].

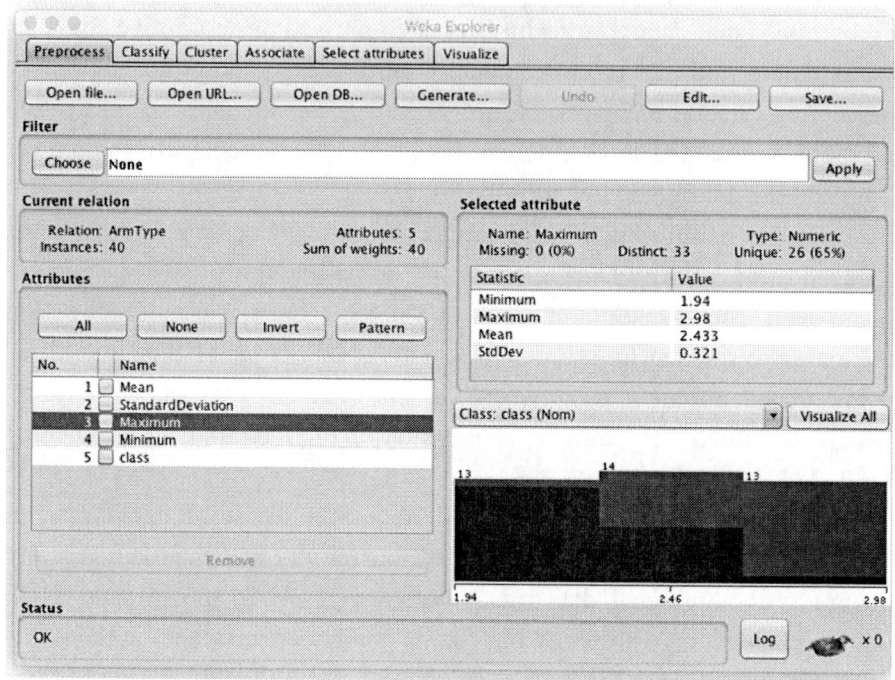

Figure 6. The WEKA Explorer for deriving machine learning classification accuracy to differentiate reduced arm swing for a subject with Erb's palsy [82].

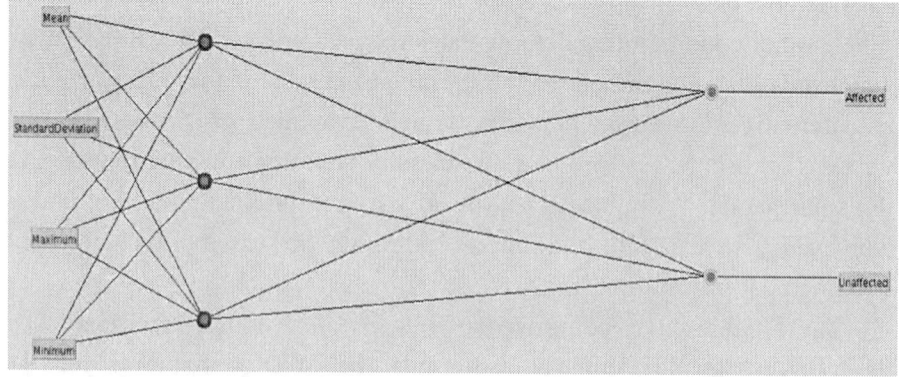

Figure 7. The multilayer perceptron neural network for establishing classification accuracy to distinguish reduced arm swing for a subject with Erb's palsy [82].

MULTILAYER PERCEPTRON NEURAL NETWORK FOR ASCERTAINING EFFICACY OF DEEP BRAIN STIMULATION FOR THE TREATMENT OF MOVEMENT DISORDER

The quantification of movement disorder status, such as Parkinson's disease, is highly relevant for the prescription of an effective intervention therapy [83, 84]. Traditional means for quantifying the severity of Parkinson's disease involve the application of an ordinal scale, such as through the Unified Parkinson's Disease Rating Scale (UPDRS) [83, 84, 85]. Many other similar ordinal scales exist, for which there is not an approach for translating the findings between the ordinal scales [86]. Furthermore, the reliability of these ordinal scales is a subject of controversy [87]. The wearable and wireless inertial sensor system offers considerable utility in terms of objective quantification for movement disorder tremor characteristics [4, 5, 15, 19].

Original wearable accelerometer systems demonstrated the opportunity of objectively quantifying movement disorder status for an assortment of intervention scenarios [88, 89, 90, 91, 92, 93], Intuitively, with the progressive evolution of wearable inertial sensor systems the advent of wireless applications became rampantly prevalent [94]. Wireless accelerometer systems were applied for quantifying the representation of hand tremor for Parkinson's disease [95, 96, 97].

During 2010 LeMoyne et al. applied a smartphone as a wearable and wireless accelerometer platform for quantifying Parkinson's disease hand tremor. The smartphone was worn about the dorsum of the hand through a glove. Statistically significant contrast was determined relative to a person without Parkinson's disease. The trial data was transmitted wirelessly through connectivity to the Internet, which enabled the experiment to be conducted in Pittsburgh, Pennsylvania with the post-processing occurring on the other side of the continental United States in the regional Los Angeles, California area [98]. This is considered a seminal moment for the proof of concept perspective of Network Centric Therapy [1, 2, 4, 5].

The application was extrapolated to the evaluation of deep brain stimulation efficacy for movement disorders, such as Essential Tremor. Using the smartphone mounted to the hand the efficacy of the deep brain stimulation system was ascertained through differentiating between 'On' and 'Off' status for a subject with Essential Tremor through machine learning, such as a support vector machine [99]. Further endeavors applied the multilayer perceptron neural network to differentiate between deep brain stimulation 'On' and 'Off' status using a smartphone as a wearable and wireless inertial sensor system for subjects with Parkinson's disease and Essential Tremor [100, 101, 102].

These research applications successfully demonstrate the utility of Network Centric Therapy for ascertaining the efficacy of the deep brain stimulation system. Network Centric Therapy features the amalgamation of wearable and wireless inertial sensor systems with the multilayer perceptron neural network representing the intrinsic machine learning classification resource. These preliminary research, development, testing, and evaluation endeavors establish a pathway to the ability to optimize deep brain stimulation parameter configurations in real-time with Internet connectivity bridging geographically separated experimental and post-processing resources [4, 5, 6, 9].

The next evolutionary transition is the application of the BioStamp nPoint, which is on the order of ten times lighter than the smartphone with a profile on the order of a bandage. Also, rather than requiring a glove-like mounting technique for the smartphone, the BioStamp nPoint can be simply mounted to the dorsum of the hand through an adhesive medium. Figure 8 features the Biostamp nPoint situated about the dorsum of the hand. In light of this notable evolution for wearable and wireless inertial sensor systems, the multilayer perceptron neural network remains an inherently central aspect for the acquisition of machine learning classification accuracy to distinguish between scenarios of deep brain stimulation status. Preliminary research, development, testing, and evaluation of the BioStamp nPoint for ascertaining deep brain stimulation efficacy with the multilayer perceptron neural network has been successfully conducted for the domain of Parkinson's disease [16, 17].

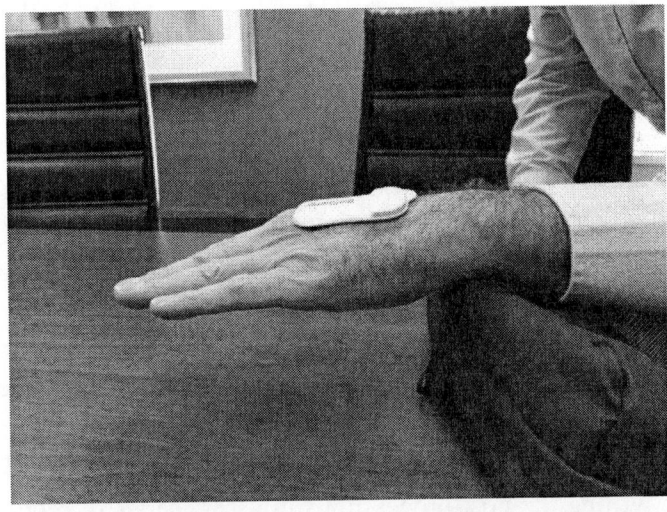

Figure 8. The BioStamp nPoint mounted to the dorsum of the hand [16, 17].

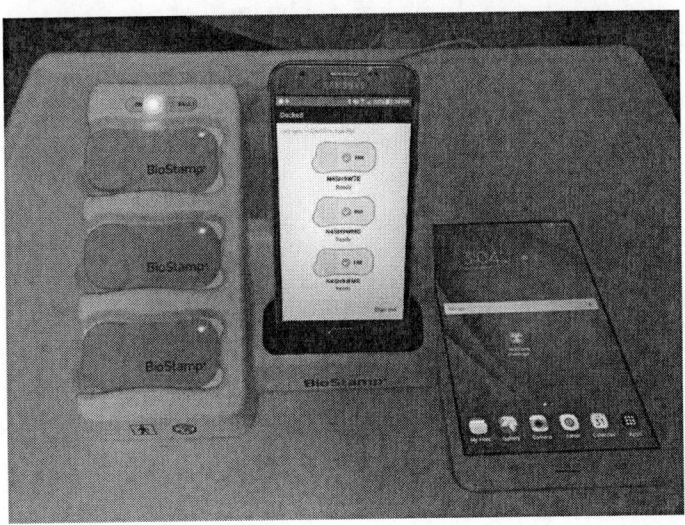

Figure 9. The BioStamp nPoint experimental apparatus [16, 17].

The operation principles of the BioStamp nPoint are representative of Network Centric Therapy. The BioStamp nPoint achieves local wireless connectivity with a tablet for operation. The associated apparatus is illustrated in Figure 9. After the acquisition of an experimental trial the signal data is transmitted in conjunction with a smartphone to a secure

Cloud computing environment. The BioStamp nPoint is an FDA 510(k) cleared medical device that is suitable for the acquisition of medical grade data [16, 17, 103].

The multilayer perceptron neural network has been successfully applied as a machine learning algorithm for the research objective of differentiating between deep brain stimulation set to 'On' and 'Off' status for a subject with Parkinson's disease. The accelerometer signal data acquired by the BioStamp nPoint quantifies the status of the hand tremor. Upon the transmission of the accelerometer signal data to the Cloud computing environment the next phase of the experimental process is to post-process the signal data into a feature set for machine learning classification [16, 17].

Before writing the post-processing software a series of procedures are recommended to clearly communicate the software development strategy. For example a series of requirements can be instilled:

- Read the data files.
- Synthesize the loaded data into acceleration magnitude as a function of time.
- Graphically visualize the acceleration magnitude as a function of time.
- Post-process the acceleration magnitude to numeric values suitable for the feature set.
- Write the feature set to an Attribute-Relation File Format (ARFF) for the Waikato Environment for Knowledge Analysis (WEKA) [17].

Although these requirements may appear simple for achieving the research objective from an engineering proof of concept perspective, more sophisticated endeavors for data science benefit from a documentation of the software development thought process. The Fagan inspection can also be incorporated. The Fagan inspection applies a thorough review of the requirements for their clarity, appropriateness, and relevance [104, 105].

With the establishment of requirements the programming language can be selected. A prevalent programming language for the domain of data science is Python [106, 107]. Python has been selected as the programming language to consolidate the accelerometer signal data into a feature set amenable for WEKA [17].

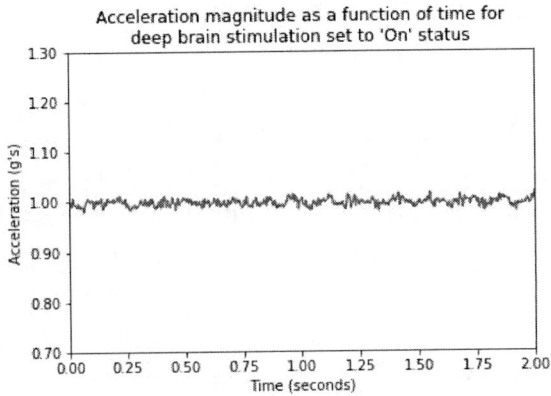

Figure 10. Accelerometer magnitude signal data for Parkinson's disease hand tremor with deep brain stimulation system set to 'On' status [16, 17].

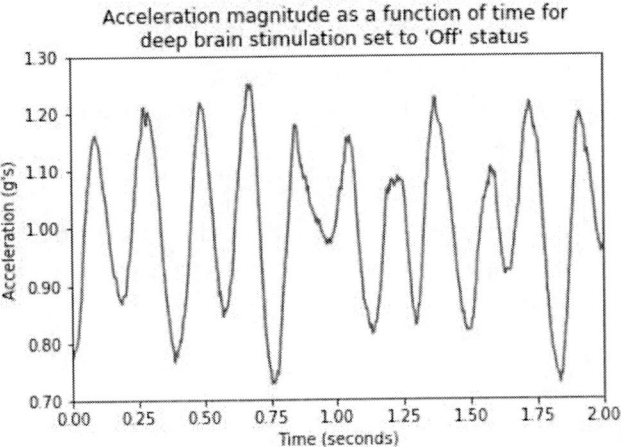

Figure 11. Accelerometer magnitude signal data for Parkinson's disease hand tremor with deep brain stimulation system set to 'Off' status [16, 17].

Further refinement of the requirements can lead to a pseudo-code, which is a series of instructions that satisfies the requirements and provides a clear set of instructions that the software program is envisioned to achieve. This phase of the software development process bridges the conceptual to the applied. With the pseudo-code developed the syntax can be established for the software program [107, 108].

The visualized data of the acceleration magnitude ascertains notable disparity of the subject with Parkinson's disease for the deep brain stimulation system set to 'On' and 'Off'. Figure 10 provides a graphical representation of the Parkinson's disease hand tremor for the subject with deep brain stimulation set to 'On' status. The subject displays considerably greater hand tremor amplitude with the deep brain stimulation system set to 'Off' status as shown in Figure 11 [16, 17].

The post-processing software using Python consolidates the accelerometer magnitude signal data into an ARFF for machine learning classification using WEKA. The feature set for the ARFF is composed of five numeric attributes:

- Maximum of the acceleration magnitude signal
- Minimum of the acceleration magnitude signal
- Mean of the acceleration magnitude signal
- Standard deviation of the acceleration magnitude signal
- Coefficient of variation of the acceleration magnitude signal

These numeric attributes are the basis for the machine learning algorithm to distinguish between the two established classes, which are 'On' status and 'Off' status for deep brain stimulation [16, 17].

The multilayer perceptron neural network implemented through WEKA attained considerable classification accuracy for distinguishing between deep brain stimulation set to 'On' and 'Off' status for a subject with Parkinson's disease. Figure 12 provides the WEKA Explorer with the uploaded ARFF. Figure 13 represents the multilayer perceptron neural network generated by WEKA to achieve the machine learning classification model [16, 17].

Machine Learning Classification for Network Centric Therapy ... 59

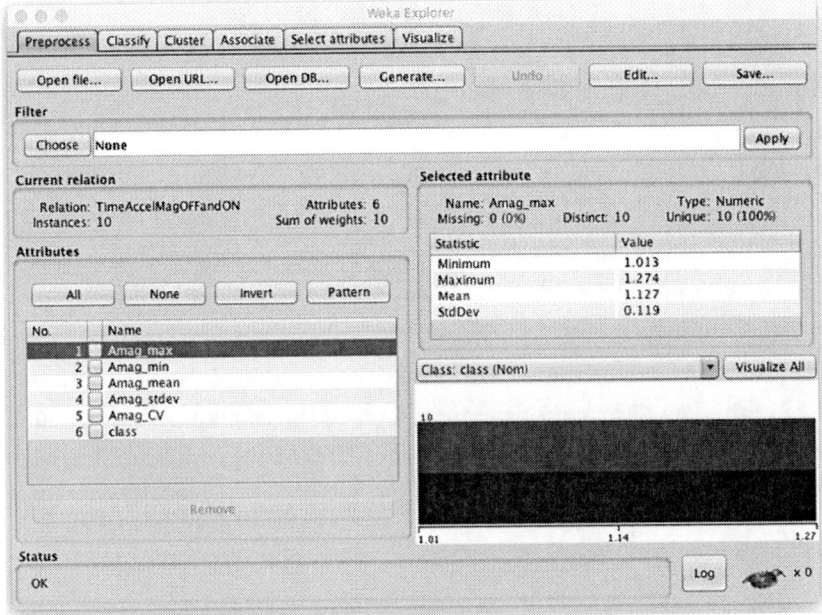

Figure 12. The WEKA Explorer with the uploaded ARFF for establishing machine learning classification between deep brain stimulation set to 'On' and 'Off' status for a subject with Parkinson's disease [16, 17].

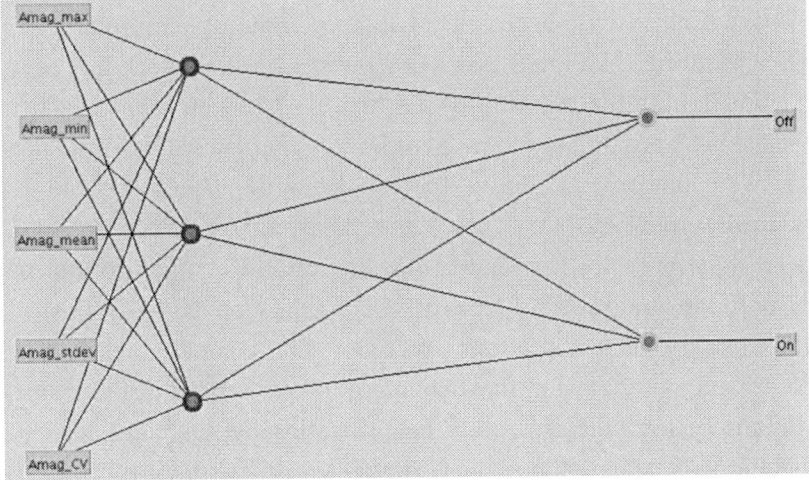

Figure 13. The multilayer perceptron neural network attaining machine learning classification accuracy between deep brain stimulation set to 'On' and 'Off' status for a subject with Parkinson's disease [16, 17].

FUTURE PERSPECTIVES

In the near future the real-time optimization of the deep brain stimulation parameter configuration through Network Centric Therapy is envisioned. Foundational is the inherent utility of the multilayer perceptron neural network. Wearable and wireless systems are foreseen to undergo considerable evolution. Network Centric Therapy is predicted to provide clinicians with advanced situational awareness of patient health status with substantial improvements for diagnostic and prognostic acuity. Network Centric Therapy enables a pathway for the confluence of the Internet of Things with Data Science for the treatment of movement disorders, such as Parkinson's disease and Essential tremor. Transition to deep learning is predicted to impact the opportunities provided by Network Centric Therapy.

CONCLUSION

Network Centric Therapy encompasses the confluence of wearable and wireless systems with access to Cloud computing environments and machine learning. The multilayer perceptron neural network has been the foundational machine learning algorithm for Network Centric Therapy. This machine learning algorithm incorporates characteristics of the neuron, which is a fundamental basis for the brain's perceptive acuity. The Waikato Environment for Knowledge Analysis (WEKA) provides the software platform for implementing the multilayer perceptron neural network for an assortment of Network Centric Therapy applications.

Preliminary demonstrations incorporated wearable and wireless accelerometer systems for the quantification of tendon reflex response. Evolutions of the wireless reflex quantification system have lead to the application of smartphones and portable media devices, which are equipped with accelerometer and gyroscope sensors, capable of accessing a functional Cloud computing environment through email. With the incorporation of machine learning, such as the multilayer perceptron neural network, differentiation of a hemiplegic reflex pair has been achieved with considerable classification accuracy. These capabilities have been extrapolated to attaining considerable classification accuracy to distinguish between an affected and unaffected lower limb during hemiplegic gait.

The smartphone has been demonstrated for quantifying tremor status for both Parkinson's disease and Essential tremor. With the multilayer perceptron neural network and the smartphone as a wearable and wireless inertial sensor platform utilizing access to a functional Cloud computing environment considerable classification accuracy has been attained to distinguish between deep brain stimulation set to 'On' and 'Off' status. These achievements demonstrate preliminary proof of concept for Network Centric Therapy.

Further development of Network Centric Therapy with the multilayer perceptron neural network has involved the BioStamp nPoint for the treatment of Parkinson's disease through the application of a deep brain stimulation system. The BioStamp nPoint is an FDA 510(k) cleared medical device that provides the acquisition of medical grade data with access to a secure Cloud computing environment. The signal data has been consolidated to a feature set through software automation using Python with a basic discussion of the software development process. The multilayer perceptron neural network achieved considerable classification accuracy to distinguish between the deep brain stimulation system set to 'On' and 'Off' status for Parkinson's disease hand tremor objectively quantified by the BioStamp nPoint as a highly wearable and wireless inertial sensor system. In summary, these developments implicate the considerable influence that Network Centric Therapy is envisioned to have for the biomedical engineering and healthcare industries.

Future evolutions of Network Centric Therapy with the synergy of wearable and wireless systems accessible to Cloud computing environments equipped with machine learning, such as the multilayer perceptron neural network, are envisioned. A readily feasible endeavor is the real-time optimization of parameter configurations for deep brain stimulation systems treating movement disorders, such as Parkinson's disease and Essential tremor. These capabilities are predicted to substantially advance clinical situational awareness, diagnostic, and prognostic acuity. The advent of Network Centric Therapy with the multilayer perceptron neural network and associated machine learning strategies are forecasted to substantially advance biomedical engineering and healthcare industries in the context of data science in conjunction with the Internet of Things manifested by wearable and wireless inertial sensor systems.

REFERENCES

[1] LeMoyne, Robert, and Timothy Mastroianni. 2018. *Wearable and Wireless Systems for Healthcare I: Gait and Reflex Response Quantification*. Singapore: Springer.

[2] LeMoyne, Robert, and Timothy Mastroianni. 2018. "Wearable and Wireless Systems for Gait Analysis and Reflex Quantification." In *Wearable and Wireless Systems for Healthcare I: Gait and Reflex Response Quantification*, 1-20. Singapore: Springer.

[3] LeMoyne, Robert, and Timothy Mastroianni. 2018. "Future Perspective of Network Centric Therapy." In *Wearable and Wireless Systems for Healthcare I: Gait and Reflex Response Quantification*, 133-134. Singapore: Springer.

[4] LeMoyne, Robert, Timothy Mastroianni, Donald Whiting, and Nestor Tomycz. 2019. *Wearable and Wireless Systems for Healthcare II: Movement Disorder Evaluation and Deep Brain Stimulation Systems*. Singapore: Springer.

[5] LeMoyne, Robert, Timothy Mastroianni, Donald Whiting, and Nestor Tomycz. 2019. "Wearable and Wireless Systems for Movement Disorder Evaluation and Deep Brain Stimulation Systems." In *Wearable and Wireless Systems for Healthcare II: Movement Disorder Evaluation and Deep Brain Stimulation Systems*, 1-15. Singapore: Springer.

[6] LeMoyne, Robert, Timothy Mastroianni, Donald Whiting, and Nestor Tomycz. 2019. "New Perspectives for Network Centric

Therapy for the Treatment of Parkinson's Disease and Essential Tremor." In *Wearable and Wireless Systems for Healthcare II: Movement Disorder Evaluation and Deep Brain Stimulation Systems*, 127-128. Singapore: Springer.

[7] LeMoyne, Robert, Cristian Coroian, Timothy Mastroianni, Pawel Opalinski, Michael Cozza, and Warren Grundfest. 2009. "The Merits of Artificial Proprioception, with Applications In *Biofeedback Gait Rehabilitation Concepts and Movement Disorder Characterization*." In *Biomedical Engineering*, edited by C. A. Barros de Mello, 165–198. Vienna: InTech.

[8] LeMoyne, Robert, Cristian Coroian, Timothy Mastroianni, and Warren Grundfest. 2008. "Accelerometers for quantification of gait and movement disorders: a perspective review." *Journal of Mechanics in Medicine and Biology* 8:137-152.

[9] LeMoyne, Robert, and Timothy Mastroianni. 2019. "Network Centric Therapy for Wearable and Wireless Systems." In *Smartphones: Recent Innovations and Applications*, edited by Paolo Dabove, Chapter 7. Hauppauge, New York: Nova Science Publishers.

[10] LeMoyne, Robert, and Timothy Mastroianni. 2015. "Use of smartphones and Portable Media Devices for Quantifying Human Movement Characteristics of Gait, Tendon Reflex Response, and Parkinson's Disease Hand Tremor." In *Mobile Health Technologies: Methods and Protocols*, edited by A. Rasooly and K. E. Herold, 335-358. New York: Springer.

[11] LeMoyne, Robert, and Timothy Mastroianni. 2017. "Wearable and Wireless Gait Analysis Platforms: Smartphones and Portable Media Devices." In *Wireless MEMS Networks and Applications*, edited by D. Uttamchandani, 129-152. New York: Elsevier.

[12] LeMoyne, Robert, and Timothy Mastroianni. 2016. "Telemedicine Perspectives for Wearable and Wireless Applications Serving the Domain of Neurorehabilitation and Movement Disorder Treatment." In *Telemedicine*, 1-10. Dover, Delaware: SMGroup.

[13] LeMoyne, Robert, and Timothy Mastroianni. 2017. "Smartphone and Portable Media Device: A Novel Pathway Toward the Diagnostic Characterization of Human Movement." In *Smartphones from an Applied Research Perspective*, edited by N. Mohamudally, Rijeka, Croatia: InTech.

[14] LeMoyne, Robert, and Timothy Mastroianni. 2018. "Smartphones and Portable Media Devices as Wearable and Wireless Systems for Gait and Reflex Response Quantification." In *Wearable and Wireless Systems for Healthcare I: Gait and Reflex Response Quantification*, 73-93. Singapore: Springer.

[15] LeMoyne, Robert, Timothy Mastroianni, Donald Whiting, and Nestor Tomycz. 2019. "Wearable and Wireless Systems with Internet Connectivity for Quantification of Parkinson's Disease and Essential Tremor Characteristics." In *Wearable and Wireless Systems for Healthcare II: Movement Disorder Evaluation and Deep Brain Stimulation Systems*, 79-97. Singapore: Springer.

[16] LeMoyne, Robert, Timothy Mastroianni, Donald Whiting, and Nestor Tomycz. 2019. "Network Centric Therapy for deep brain stimulation status parametric analysis with machine learning classification." Presented at *49th Society for Neuroscience Annual Meeting* (Nanosymposium), Chicago, Illinois, October 19-23.

[17] LeMoyne, Robert, Timothy Mastroianni, Donald Whiting, and Nestor Tomycz. 2019. "Preliminary Network Centric Therapy for Machine Learning Classification of Deep Brain Stimulation Status for the Treatment of Parkinson's Disease with a Conformal Wearable and Wireless Inertial Sensor." *Advances in Parkinson's Disease* 8: 75-91.

[18] LeMoyne, Robert, and Timothy Mastroianni. 2018. "Role of Machine Learning for Gait and Reflex Response Classification." In *Wearable and Wireless Systems for Healthcare I: Gait and Reflex Response Quantification*, 111-120. Singapore: Springer.

[19] LeMoyne, Robert, Timothy Mastroianni, Donald Whiting, and Nestor Tomycz. 2019. "Role of Machine Learning for Classification of Movement Disorder and Deep Brain Stimulation Status." In

Wearable and Wireless Systems for Healthcare II: Movement Disorder Evaluation and Deep Brain Stimulation Systems, 99-111. Singapore: Springer.

[20] LeMoyne, Robert. 2016. "Testing and Evaluation Strategies for the Powered Prosthesis, a Global Perspective." In *Advances for Prosthetic Technology: From Historical Perspective to Current Status to Future Application*, 37-58. Tokyo: Springer.

[21] Munakata, Toshinori. 2008. *Fundamentals of the New Artificial Intelligence: Neural, Evolutionary, Fuzzy and More*. London: Springer.

[22] Kandel, Eric R., James H. Schwartz, and Thomas M. Jessell. 2000. *Principles of Neural Science*. New York: McGraw-Hill.

[23] Seeley, Rod R., Trent D. Stephens, and Phillip Tate. 2003. *Anatomy and Physiology*, Boston: McGraw-Hill.

[24] Hall, Mark, Eibe Frank, Geoffrey Holmes, Bernhard Pfahringer, Peter Reutemann, and Ian H. Witten. 2009. "The WEKA data mining software: an update." *ACM SIGKDD Explorations Newsletter* 11:10-18.

[25] Witten, Ian H., Eibe Frank, and Mark A. Hall. 2011. *Data Mining: Practical Machine Learning Tools and Techniques*. Burlington, Massachusetts: Morgan Kaufmann.

[26] WEKA [http://www.cs.waikato.ac.nz/~ml/weka/]

[27] Witten, Ian H., Eibe Frank, and Mark A. Hall. 2011. "Implementation: Real Machine Learning Schemes." In *Data Mining: Practical Machine Learning Tools and Techniques*, 191-304. Burlington, Massachusetts: Morgan Kaufmann.

[28] Russell, Stuart and Peter Norvig. 2010. "Learning from Examples." In *Artificial Intelligence: A Modern Approach*, 693-767. New York: Prentice Hall.

[29] Bryson, Arthur E. and Yu-Chi Ho. 1969. *Applied Optimal Control*. Waltham, Massachusetts: Blaisdell.

[30] Russell, Stuart and Peter Norvig. 2010. "Introduction." In *Artificial Intelligence: A Modern Approach*, 1-33. New York: Prentice Hall.

[31] LeMoyne, Robert, Timothy Mastroianni, Donald Whiting, and Nestor Tomycz. 2019. "Assessment of Machine Learning Classification Strategies for the Differentiation of Deep Brain Stimulation "On" and "Off" Status for Parkinson's Disease Using a Smartphone as a Wearable and Wireless Inertial Sensor for Quantified Feedback." In *Wearable and Wireless Systems for Healthcare II: Movement Disorder Evaluation and Deep Brain Stimulation Systems*, 113-126. Singapore: Springer.

[32] LeMoyne, Robert, Timothy Mastroianni, Cyrus McCandless, Christopher Currivan, Donald Whiting, and Nestor Tomycz. 2018. "Implementation of a smartphone as a wearable and wireless inertial sensor platform for determining efficacy of deep brain stimulation for Parkinson's disease tremor through machine learning." Presented at *48th Society for Neuroscience Annual Meeting*, (Nano-symposium), San Diego, California, November 3-7.

[33] LeMoyne, Robert, Timothy Mastroianni, Cristian Coroian, and Warren Grundfest. 2011. "Tendon reflex and strategies for quantification, with novel methods incorporating wireless accelerometer reflex quantification devices, a perspective review." *Journal of Mechanics in Medicine and Biology* 11:471-513.

[34] Bickley, Lynn, and Peter G. Szilagyi. 2003. *Bates' guide to physical examination and history-taking*. Philadelphia: Lippincott Williams & Wilkins.

[35] Litvan, I., C. A. Mangone, W. Werden, J. A. Bueri, C. J. Estol, D. O. Garcea, R. C. Rey, R. E. P. Sica, M. Hallett, and J. J. Bartko. 1996. "Reliability of the NINDS myotatic reflex scale." *Neurology* 47:969-972.

[36] Manschot, S., L. Van Passel, E. Buskens, A. Algra, and J. Van Gijn. 1998. "Mayo and NINDS scales for assessment of tendon reflexes: between observer agreement and implications for communication." *Journal of Neurology, Neurosurgery & Psychiatry* 64:253-255.

[37] Stam, J., and H. Van Crevel. 1990. "Reliability of the clinical and electromyographic examination of tendon reflexes." *Journal of Neurology* 237:427-431.

[38] Van de Crommert, H. W. A. A., M. Faist, W. Berger, and J. Duysens. 1996. "Biceps femoris tendon jerk reflexes are enhanced at the end of the swing phase in humans." *Brain Research* 734:341-344.

[39] Faist, Michael, Matthias Ertel, Wiltrud Berger, and Volker Dietz. 1999. "Impaired modulation of quadriceps tendon jerk reflex during spastic gait: differences between spinal and cerebral lesions." *Brain* 122:567-579.

[40] Cozens, J. Alastair, Simon Miller, Iain R. Chambers, and A. David Mendelow. 2000. "Monitoring of head injury by myotatic reflex evaluation." *Journal of Neurology, Neurosurgery & Psychiatry* 68:581-588.

[41] Pagliaro, Paolo, and Paola Zamparo. 1999. "Quantitative evaluation of the stretch reflex before and after hydro kinesy therapy in patients affected by spastic paresis." *Journal of Electromyography and Kinesiology* 9:141-148.

[42] Zhang, Li-Qun, Guangzhi Wang, Takashi Nishida, Dali Xu, James A. Sliwa, and W. Zev Rymer. 2000. "Hyperactive tendon reflexes in spastic multiple sclerosis: measures and mechanisms of action." *Archives of Physical Medicine and Rehabilitation* 81:901-909.

[43] Koceja, David M., and Gary Kamen. 1988. "Conditioned patellar tendon reflexes in sprint-and endurance-trained athletes." *Medicine and Science in Sports and Exercise* 20:172-177.

[44] Kamen, Gary, and David M. Koceja. 1989. "Contralateral influences on patellar tendon reflexes in young and old adults." *Neurobiology of Aging* 10:311-315.

[45] Lebiedowska, Maria K., and John Robert Fisk. 2003. "Quantitative evaluation of reflex and voluntary activity in children with spasticity1." *Archives of Physical Medicine and Rehabilitation* 84:828-837.

[46] Mamizuka, Naotaka, Masataka Sakane, Koji Kaneoka, Noriyuki Hori, and Naoyuki Ochiai. 2007. "Kinematic quantitation of the patellar tendon reflex using a tri-axial accelerometer." *Journal of Biomechanics* 40:2107-2111.

[47] Tham, Lai Kuan, Noor Azuan Abu Osman, Wan Abu Bakar Wan Abas, and Kheng Seang Lim. 2013. "The validity and reliability of motion analysis in patellar tendon reflex assessment." *PLoS One* 8:e55702.

[48] Tham, L. K., NA Abu Osman, K. S. Lim, B. Pingguan-Murphy, WAB Wan Abas, and N. Mohd Zain. 2011. "Investigation to predict patellar tendon reflex using motion analysis technique." *Medical Engineering & Physics* 33:407-410.

[49] Chandrasekhar, Annapoorna, Noor Azuan Abu Osman, Lai Kuan Tham, Kheng Seang Lim, and Wan Abu Bakar Wan Abas. 2013. "Influence of age on patellar tendon reflex response." *PloS One* 8:e80799.

[50] LeMoyne, Robert Charles. 2010. *Wireless Quantified Reflex Device*. University of California, Los Angeles, Ph.D. Dissertation.

[51] LeMoyne, Robert, Roozbeh Jafari, and David Jea. 2005. "Fully quantified evaluation of myotatic stretch reflex." Presented at *35th Society for Neuroscience Annual Meeting*, Washington D.C., November 12-16.

[52] LeMoyne, Robert, Foad Dabiri, and Roozbeh Jafari. 2008. "Quantified deep tendon reflex device, second generation." *Journal of Mechanics in Medicine and Biology* 8:75-85.

[53] LeMoyne, Robert, Cristian Coroian, Timothy Mastroianni, and Warren Grundfest. 2008. "Quantified deep tendon reflex device for response and latency, third generation." *Journal of Mechanics in Medicine and Biology* 8:491-506.

[54] LeMoyne, Robert, Cristian Coroian, and Timothy Mastroianni. 2009. "Wireless accelerometer reflex quantification system characterizing response and latency." In Engineering in Medicine and Biology Society (EMBC), *31st Annual International Conference of the IEEE, Minneapolis*, Minnesota, September 3-6.

[55] LeMoyne, Robert, Timothy Mastroianni, Halo Kale, Jorge Luna, Joshua Stewart, Stephen Elliot, Filip Bryan, Cristian Coroian, and Warren Grundfest. 2011. "Fourth generation wireless reflex

quantification system for acquiring tendon reflex response and latency." *Journal of Mechanics in Medicine and Biology* 11:31-54.

[56] LeMoyne, Robert, Cristian Coroian, and Timothy Mastroianni. 2009. "Evaluation of a wireless three dimensional MEMS accelerometer reflex quantification device using an artificial reflex system." In *International Conference on Complex Medical Engineering (ICME) of the IEEE*, Tempe, Arizona, April 9-11.

[57] LeMoyne, Robert, Timothy Mastroianni, Cristian Coroian, and Warren Grundfest. 2010. "Wireless three dimensional accelerometer reflex quantification device with artificial reflex system." *Journal of Mechanics in Medicine and Biology* 10:401-415.

[58] LeMoyne, Robert, Cristian Coroian, and Timothy Mastroianni. 2009. "Wireless reflex quantification system classifying frequency domain analysis disparity of deep tendon reflex." Presented at *39th Society for Neuroscience Annual Meeting*, Chicago, Illinois, October 17-21.

[59] LeMoyne, Robert, Cristian Coroian, and Timothy Mastroianni. 2009. "Consideration of the frequency domain as a strategy for classifying tendon reflex response." Presented at *UCLA Brain Research Institute 2009 Neuroscience Poster Session* Los Angeles, California, November.

[60] LeMoyne, Robert, Cristian Coroian, Timothy Mastroianni, Michael Cozza, and Warren Grundfest. 2010. "Quantification of reflex response through an iPhone wireless accelerometer application." Presented at *40th Society for Neuroscience Annual Meeting*, San Diego, California, November 13-17.

[61] LeMoyne, Robert and Timothy Mastroianni. 2011. "Reflex response quantification using an iPod wireless accelerometer application." Presented at *41st Society for Neuroscience Annual Meeting*, (featured in Hot Topics; top 1% of abstracts), Washington D.C., November 12-16.

[62] LeMoyne, Robert, Timothy Mastroianni, and Warren Grundfest. 2012. "Quantified reflex strategy using an iPod as a wireless accelerometer application." In Engineering in Medicine and Biology

Society (EMBC), *34th Annual International Conference of the IEEE*, San Diego, California, August 28-September 1.

[63] LeMoyne, Robert, Timothy Mastroianni, Warren Grundfest, and Kiisa Nishikawa. 2013. "Implementation of an iPhone wireless accelerometer application for the quantification of reflex response." In Engineering in Medicine and Biology Society (EMBC), *35th Annual International Conference of the IEEE*, Osaka, Japan, July 3-7.

[64] LeMoyne, Robert, Wesley T. Kerr, Kevin Zanjani, and Timothy Mastroianni. 2014. "Implementation of an iPod wireless accelerometer application using machine learning to classify disparity of hemiplegic and healthy patellar tendon reflex pair." *Journal of Medical Imaging and Health Informatics* 4:21-28.

[65] LeMoyne, Robert, and Timothy Mastroianni. 2014. "Implementation of a smartphone as a wireless gyroscope application for the quantification of reflex response." In Engineering in Medicine and Biology Society (EMBC), *36th Annual International Conference of the IEEE*, Chicago, Illinois, August 26-30.

[66] LeMoyne, Robert, and Timothy Mastroianni. 2014. "Quantification of patellar tendon reflex response using an iPod wireless gyroscope application with experimentation conducted in Lhasa, Tibet and post-processing conducted in Flagstaff, Arizona through wireless Internet connectivity." Presented at *44th Society for Neuroscience Annual Meeting*, Washington D.C., November 15-19.

[67] LeMoyne, Robert, and Timothy Mastroianni. 2016. "Smartphone wireless gyroscope platform for machine learning classification of hemiplegic patellar tendon reflex pair disparity through a multilayer perceptron neural network." In *Wireless Health conference of the IEEE*, Bethesda, MD, October 25-27.

[68] LeMoyne, Robert, and Timothy Mastroianni. 2016. "Implementation of a multilayer perceptron neural network for classifying a hemiplegic and healthy reflex pair using an iPod wireless gyroscope platform." Presented at *46th Society for Neuroscience Annual Meeting*, San Diego, California, November 12-16.

[69] LeMoyne, Robert, Cristian Coroian, Timothy Mastroianni, and Warren Grundfest. 2008. "Virtual proprioception." *Journal of Mechanics in Medicine and Biology* 8:317-338.

[70] LeMoyne, Robert, Cristian Coroian, Timothy Mastroianni, and Warren Grundfest. 2009. "Wireless accelerometer assessment of gait for quantified disparity of hemiparetic locomotion." *Journal of Mechanics in Medicine and Biology* 9:329-343.

[71] LeMoyne, Robert, Timothy Mastroianni, Michael Cozza, Cristian Coroian, and Warren Grundfest. 2010. "Implementation of an iPhone as a wireless accelerometer for quantifying gait characteristics." In Engineering in Medicine and Biology Society (EMBC), *32nd Annual International Conference of the IEEE*, Buenos Aires, Argentina, August 31-September 4.

[72] LeMoyne, Robert, Timothy Mastroianni, Michael Cozza, and Cristian Coroian. 2010. "Quantification of gait characteristics through a functional iPhone wireless accelerometer application mounted to the spine." In *ASME 5th Frontiers in Biomedical Devices Conference*, Newport Beach, California, September 20–21.

[73] LeMoyne, Robert, Timothy Mastroianni, Michael Cozza, and Cristian Coroian. 2010. "iPhone wireless accelerometer application for acquiring quantified gait attributes." In *ASME 5th Frontiers in Biomedical Devices Conference*, Newport Beach, California, September 20–21.

[74] Mellone, Sabato, Carlo Tacconi, and Lorenzo Chiari. 2012. "Validity of a Smartphone-based instrumented Timed Up and Go." *Gait & Posture* 36:163-165.

[75] Capela, Nicole A., Edward D. Lemaire, and Natalie Baddour. 2015. "Novel algorithm for a smartphone-based 6-minute walk test application: algorithm, application development, and evaluation." *Journal of Neuroengineering and Rehabilitation* 12:1-13.

[76] Raknim, Paweeya, and Kun-chan Lan. 2016. "Gait monitoring for early neurological disorder detection using sensors in a smartphone: Validation and a case study of parkinsonism." *Telemedicine and e-Health* 22:75-81.

[77] LeMoyne, Robert, Timothy Mastroianni, and Warren Grundfest. 2011. "Wireless accelerometer iPod application for quantifying gait characteristics." In Engineering in Medicine and Biology Society (EMBC), *33rd Annual International Conference of the IEEE*, Boston, Massachusetts, August 30-September 3.

[78] LeMoyne, Robert, and Timothy Mastroianni. 2014. "Implementation of an iPod application as a wearable and wireless accelerometer system for identifying quantified disparity of hemiplegic gait." *Journal of Medical Imaging and Health Informatics* 4:634-641.

[79] LeMoyne, Robert, and Timothy Mastroianni. 2018. "Implementation of a smartphone as a wireless accelerometer platform for quantifying hemiplegic gait disparity in a functionally autonomous context." *Journal of Mechanics in Medicine and Biology* 18:1850005.

[80] LeMoyne, Robert, and Timothy Mastroianni. 2018. "Implementation of a smartphone as a wearable and wireless gyroscope platform for machine learning classification of hemiplegic gait through a multilayer perceptron neural network." In *17th International Conference on Machine Learning and Applications (ICMLA) of the IEEE*, Orlando, Florida, December 17-20.

[81] LeMoyne, Robert, and Timothy Mastroianni. 2016. "Implementation of a smartphone as a wireless gyroscope platform for quantifying reduced arm swing in hemiplegic gait with machine learning classification by multilayer perceptron neural network." In Engineering in Medicine and Biology Society (EMBC), *38th Annual International Conference of the IEEE,* Orlando, Florida, August 16-20.

[82] Mastroianni, Timothy and Robert LeMoyne. 2016. "Application of a multilayer perceptron neural network with an iPod as a wireless gyroscope platform to classify reduced arm swing gait for people with Erb's palsy." Presented at *46th Society for Neuroscience Annual Meeting*, San Diego, California, November 12-16.

[83] LeMoyne, Robert, Timothy Mastroianni, Donald Whiting, and Nestor Tomycz. 2019. "Movement Disorders: Parkinson's Disease and Essential Tremor—A General Perspective." In *Wearable and*

Wireless Systems for Healthcare II: Movement Disorder Evaluation and Deep Brain Stimulation Systems, 17-24. Singapore: Springer.

[84] LeMoyne, Robert, Timothy Mastroianni, Donald Whiting, and Nestor Tomycz. 2019. "Traditional Ordinal Strategies for Establishing the Severity and Status of Movement Disorders, such as Parkinson's Disease and Essential Tremor." In *Wearable and Wireless Systems for Healthcare II: Movement Disorder Evaluation and Deep Brain Stimulation Systems*, 25-36. Singapore: Springer.

[85] Goetz, Christopher G., Barbara C. Tilley, Stephanie R. Shaftman, Glenn T. Stebbins, Stanley Fahn, Pablo Martinez-Martin, Werner Poewe, Cristina Sampaio, Matthew B. Stern, Richard Dodel, Bruno Dubois, Robert Holloway, Joseph Jankovic, Jaime Kulisevsky, Anthony E. Lang, Andrew Lees, Sue Leurgans, Peter A. LeWitt, David Nyenhuis, C. Warren Olanow, Olivier Rascol, Anette Schrag, Jeanne A. Teresi, Jacobus J. van Hilten, and Nancy LaPelle. 2008. "Movement Disorder Society-sponsored revision of the Unified Parkinson's Disease Rating Scale (MDS-UPDRS): Scale presentation and clinimetric testing results." *Movement Disorders* 23:2129-2170.

[86] Ramaker, Claudia, Johan Marinus, Anne Margarethe Stiggelbout, and Bob Johannes Van Hilten. 2002. "Systematic evaluation of rating scales for impairment and disability in Parkinson's disease." *Movement Disorders* 17:867-876.

[87] Post, Bart, Maruschka P. Merkus, Rob MA de Bie, Rob J. de Haan, and Johannes D. Speelman. 2005. "Unified Parkinson's disease rating scale motor examination: are ratings of nurses, residents in neurology, and movement disorders specialists interchangeable?" *Movement Disorders* 20:1577-1584.

[88] Schrag, Anette, Ludwig Schelosky, Udo Scholz, and Werner Poewe. 1999. "Reduction of parkinsonian signs in patients with Parkinson's disease by dopaminergic versus anticholinergic single-dose challenges." *Movement Disorders* 14:252-255.

[89] Keijsers, N. L. W., M. W. I. M. Horstink, J. J. Van Hilten, J. I. Hoff, and C. C. A. M. Gielen. 2000. "Detection and assessment of the

severity of Levodopa-induced dyskinesia in patients with Parkinson's disease by neural networks." *Movement Disorders* 15:1104-1111.

[90] Keijsers, Noël LW, Martin WIM Horstink, and Stan CAM Gielen. 2006. "Ambulatory motor assessment in Parkinson's disease." *Movement Disorders* 21:34-44.

[91] Gurevich, Tanya Y., Herzel Shabtai, Amos D. Korczyn, Ely S. Simon, and Nir Giladi. 2006. "Effect of rivastigmine on tremor in patients with Parkinson's disease and dementia." *Movement Disorders* 21:1663-1666.

[92] Obwegeser, Alois A., Ryan J. Uitti, Robert J. Witte, John A. Lucas, Margaret F. Turk, and Robert E. Wharen Jr. 2001. "Quantitative and qualitative outcome measures after thalamic deep brain stimulation to treat disabling tremors." *Neurosurgery* 48:274-284.

[93] Kumru, Hatice, Christopher Summerfield, Francesc Valldeoriola, and Josep Valls-Solé. 2004. "Effects of subthalamic nucleus stimulation on characteristics of EMG activity underlying reaction time in Parkinson's disease." *Movement Disorders* 19:94-100.

[94] Patel, Shyamal, Hyung Park, Paolo Bonato, Leighton Chan, and Mary Rodgers. 2012. "A review of wearable sensors and systems with application in rehabilitation." *Journal of Neuroengineering and Rehabilitation* 9:1-17.

[95] LeMoyne, Robert, Cristian Coroian, and Timothy Mastroianni. 2009. "Quantification of Parkinson's disease characteristics using wireless accelerometers." In *International Conference on Complex Medical Engineering (ICME) of the IEEE*, Tempe, Arizona, April 9-11.

[96] LeMoyne, Robert, Timothy Mastroianni, and Warren Grundfest. 2013. "Wireless accelerometer configuration for monitoring Parkinson's disease hand tremor." *Advances in Parkinson's Disease* 2:62-67.

[97] Giuffrida, Joseph P., David E. Riley, Brian N. Maddux, and Dustin A. Heldman. 2009. "Clinically deployable Kinesia™ technology for automated tremor assessment." *Movement Disorders* 24:723-730.

[98] LeMoyne, Robert, Timothy Mastroianni, Michael Cozza, Cristian Coroian, and Warren Grundfest. 2010. "Implementation of an iPhone for characterizing Parkinson's disease tremor through a wireless accelerometer application." In Engineering in Medicine and Biology Society (EMBC), *32nd Annual International Conference of the IEEE*, Buenos Aires, Argentina, August 31-September 4.

[99] LeMoyne, Robert, Nestor Tomycz, Timothy Mastroianni, Cyrus McCandless, Michael Cozza, and David Peduto. 2015. "Implementation of a smartphone wireless accelerometer platform for establishing deep brain stimulation treatment efficacy of essential tremor with machine learning." In Engineering in Medicine and Biology Society (EMBC), *37th Annual International Conference of the IEEE*, Milan, Italy, August 25-29.

[100] LeMoyne, Robert, Timothy Mastroianni, Nestor Tomycz, Donald Whiting, Michael Oh, Cyrus McCandless, Christopher Currivan, David Peduto. 2017. "Implementation of a multilayer perceptron neural network for classifying deep brain stimulation in 'On' and 'Off' modes through a smartphone representing a wearable and wireless sensor application." Presented at *47th Society for Neuroscience Annual Meeting*, (featured in Hot Topics; top 1% of abstracts), Washington D.C., November 11-15.

[101] LeMoyne, Robert, Timothy Mastroianni, Cyrus McCandless, Christopher Currivan, Donald Whiting, and Nestor Tomycz. 2018. "Implementation of a smartphone as a wearable and wireless accelerometer and gyroscope platform for ascertaining deep brain stimulation treatment efficacy of Parkinson's disease through machine learning classification." *Advances in Parkinson's Disease* 7:19-30.

[102] LeMoyne Robert, Timothy Mastroianni, Cyrus McCandless, Christopher Currivan, Donald Whiting, and Nestor Tomycz. 2018. "Implementation of a Smartphone as a Wearable and Wireless Inertial Sensor Platform for Determining Efficacy of Deep Brain Stimulation for Parkinson's Disease Tremor through Machine

Learning." Presented at *48th Society for Neuroscience Annual Meeting* (Nanosymposium), San Diego, California, November 3-7.

[103] MC10 Inc. [https://www.mc10inc.com/our-products#biostamp-npoint]

[104] Fagan, Michael E. 1986. "Advances in Software Inspections." *IEEE Transactions on Software Engineering* SE-12: 744-751.

[105] Fagan, Michael E. 1999. "Design and code inspections to reduce errors in program development." *IBM Systems Journal* 38: 258-287.

[106] Python [https://www.python.org/]

[107] Severance, Charles. 2013. *Python for Informatics: Exploring Information.* CreateSpace.

[108] Deitel, H. M., P. J. Deitel, T. R. Nieto, D. C. McPhie. 2001. "Control Structures: Part I." In *Perl: How to Program,* 60-93. Upper Saddle River, New Jersey: Prentice Hall.

In: Multilayer Perceptrons
Editor: Ruth Vang-Mata
ISBN: 978-1-53617-364-2
© 2020 Nova Science Publishers, Inc.

Chapter 3

AGE ESTIMATION BY USING MULTI-LAYER PERCEPTRON NEURAL NETWORK WITH IMAGE PROCESSING TECHNIQUES

Emre Avuçlu[1,*] *and Fatih Başçiftçi*[2,†]

[1]Department of Computer Technology,
Aksaray University, Aksaray, Turkey
[2]Department of Computer Engineering,
Technology Faculty, Selçuk University, Konya, Turkey

ABSTRACT

A person's identity, age or gender may need to be identified due to natural hazards such as disasters, or legal reasons such as inheritance and age manipulation.. In such cases accurate information may be asked about the identity of the person from the forensic sciences. Forensic science institutions try to make the age estimation process with different human

---

[*] Corresponding Author's Email: emreavuclu@aksaray.edu.tr.
[†] Corresponding Author's Email: basciftci@selcuk.edu.tr.

organs (example: teeth, bones etc.). These estimations are approximate estimated values. This study was conducted in order to provide the most accurate reports of forensic sciences. Forensic science makes the determination of age and gender through dental x-ray images. In this study, panoramic dental x-ray images were used to estimate age and gender. The database was created manually, with a total of 562 teeth images of 69 different dental classes. These images were first applied to image pre-processing techniques to achieve better results. After this process, the images were segmented and the feature extraction of images were made. Optional feature reduction was made. Feature vectors as a result of feature extraction process were presented as input to a multi-layered perceptron classifier. Segmentation was performed automatically and dynamically. The application was written in C # programming language. The highest 99.9% (full segment) and 100% (not full segment) classification success was achieved by using a multi-layered perceptron network.

Keywords: age estimation, panoramic graph, image processing techniques, multilayer perceptron model, backpropagation algorithm

INTRODUCTION

In order to reveal the characteristics of a living or dead person, an identification must be done initially. For many reasons it is necessary to make identification in both the living and the dead. One of the important objectives of forensic science applications is to clarify the suspicions about the identity or the events of unknown age. The age determination has an important place in determining the legal responsibilities and identification of living individuals. The age determination process is an area of study required in many branches of forensic sciences, anthropology and medicine. In the literature, length, weight, mental development, development of teeth and bones are used to determine the age of the individuals. The methods used in age estimation are grouped into 3 as radiological, morphological and histological. Radiological and morphological methods are most commonly used for age determination [1, 2]. The methods used in age estimation may vary from one society to another due to genetic characteristics and climate [3]. Since the teeth are more resistant to environmental conditions, high temperature, humidity, microbial activity and mechanical forces, they are more valuable than the bone in age estimation after death [4].

Developmental and physiological changes such as the appearance of mineralization in the teeth, the condition of the untreated teeth are used to determine age from the teeth [5]. Many of the age estimation techniques reported in the literature are based on age-related changes in the teeth [6].

Usually the remaining and undamaged tissues such as bone, teeth are used to estimate the age of unidentified cadavers [7-9]. Most of the radiographic examinations on the teeth are conducted through periapical radiographs, conventional and digital panoramic radiographs [10]. The growth status and radiological information of the teeth can be used for age determination [11]. Many studies on dental x-ray images were performed. Various segmentation and identification procedures were performed on dental x-ray images [12-20]. Age estimation was performed using different methods via dental x-ray images [21-24]. Age estimation was performed by measuring different areas of the tooth [25-30]. Age estimation was performed depending on the size of the skeleton in humans [31-33]. Age and gender were determined by using image processing and artificial intelligence techniques from dental x-ray images [34-37].

Panoramic dental x-ray images used in this study were collected from dental hospitals in different provinces and a dataset was created. Some image pre-processing techniques were applied to the teeth for better results. Images were segmented and feature extraction was performed on each sub-image divided into segments. The resulting feature vector was presented as input to the MLP neural network. In addition to age estimation, gender estimation was performed.

MATERIALS AND METHODS

The methods shown in Figure 1 were applied to the images in the database. The processed images are saved in separate folders (as trainingM1 and as trainingM2). The images obtained from these methods are transferred to the testM1 and testM2 folders for network testing. These images will be used as test data after training of the MLP network. Segmentation is dynamic and images can be split as desired. Algorithm 1. is used to set the inputs to be dynamically presented to the network.

Optionally, if the same indexed segments of all teeth are completely white or completely black during the feature extraction process, they are not presented to the network as a input. In order to execute this process, the Test function is used shown in Algorithm 2. In Algorithm 2. there are also sub functions within the function. It is shown briefly below.

The general methodology used in this study is shown in Figure 1.

Figure 2 shows the images of different teeth formed as a result of the pre-processing.

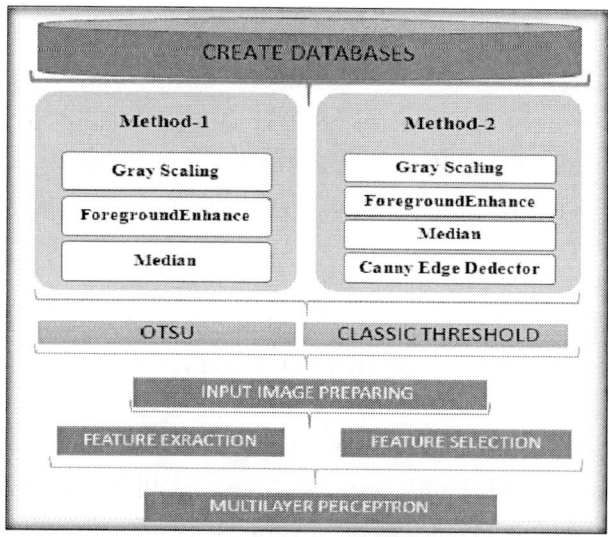

Figure 1. General metedology.

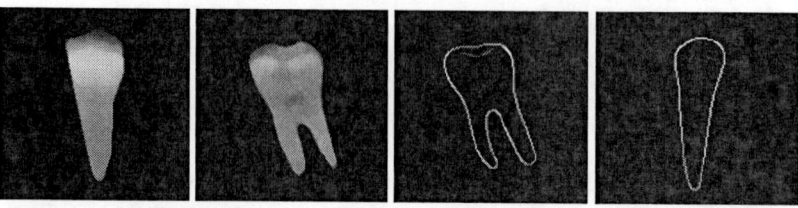

Figure 2. Pre-Process results.

Algorithm-1: Prepare inputs
1: function private void prepareInputs()
2: Inputs: scaleW, scaleH, splitW, splitH, splitSize=(scaleW * scaleH)/(splitW * splitH), dataCount= ImageSet.Count, test = [splitSize]
3: Output: result
4: for i = 0 to i < dataCount do
5: dataList ← ImageSet[i].GetCells()
6: for j = 0 to j < splitSize do
7: if (this.fullSegment) then
8: test[j] += 1
9: else
10: indx ← dataList[j].Index()
11: avgc ← dataList[j].ColorAVG()
12: if (Test(indx, avgc)) then
13: test[j] += 1
14: end if
15: end if
16: end for
17: end for
18: resultList ← List<int>()
19: for i = 0 to i < dataCount do
20: dataList ← ImageSet[i].GetCells()
21: for j = 0 to j < splitSize do
22: for r = 0 to r < resultList.Count do
23: if (j == resultList[r]) then
24: row ← (DataGridViewRow)this.dataGridInput.Rows[0].Clone()
25: age ← dataList[j].GetAge()
26: sex ← dataList[j].GetSex()
27: oms ← dataList[j].avarageabsolutedeviation()
28: row.Cells[0].Value = i + 1
29: row.Cells[1].Value = age
30: row.Cells[2].Value = sex
31: row.Cells[3].Value = j + 1
32: row.Cells[4].Value = oms
33: inputs.Add(new InputParam(i, age, sex, oms))
34: dataGridInput.Rows.Add(row)
35: end if
36: end for
37: end for
38: end for
39: inputs.CreateInputFile()
40: end function

Algorithm-2: Test function
1:function private bool Test(int indx, float clrAvg)
2:Inputs: result=false
3:Output: result
4:for i = 0 to i < ImageSet.Count do
5: imgI ← ImageSet[i].GetCells()
6: for m = 0 to m < imgI.Count do
7: if (m == indx - 1) then
8: if (clrAvg != imgI[m].ColorAVG()) then
9: result = true;
10: break;
11: end if
12: end if
13: end for
14: end for
15: return result
16:end function

Pre-Process Operations

Method

The input image is initially converted to grayscale by GrayScaling. Then the ForegroundEnhance method is applied. Then, using the Median softening filter, the resulting noise is removed after the ForegroundEnhance method. Finally, the image is converted to binary.

ForegroundEnhance

The ForegroundEnhance method was used to achieve higher success rate. Optimal segmentation of teeth with low density was achieved with this method. In this method, the procedure is as follows;

The average color between 0-255 *(i)* is calculated according to the image histogram (H_i) as in Equation 1.

$$avg = \sum_{i=0}^{255} H_i i \bigg/ \sum_{i=0}^{255} H_i \qquad (1)$$

The average color (*T*) between avg -255 is calculated according to the image histogram as in Equation 2.

$$T = \sum_{i=avg}^{255} H_i i \Big/ \sum_{i=avg}^{255} H_i \qquad (2)$$

For each pixel in the image, the comparison is made according to the *T* threshold value. Thus, the images of the teeth become more apparent as in Equation 3.

$$F[x,y] = \begin{cases} 0 & \text{if } F[x,y] > T \\ F[x,y] & \text{otherwise} \end{cases} \qquad (3)$$

Median Filter
At the same time, as a result of the color reduction process, an unwanted loud noise may appear on the new image. To remove this, use the Median filter formula as in Equation 4.

$$F[x,y] = \text{median}\{g[p,q]\} \qquad (4)$$

where *g[p, q]* refers to the convolution kernel.

In this study, a convolution matrix of 3x3 was preferred in terms of performance. In the next step, the image is converted into binary by using Otsu thresholding process according to Equation 5.

$$T[x,y] = \begin{cases} 255 & \text{if } F[x,y] \geq t \\ 0 & \text{otherwise} \end{cases} \qquad (5)$$

Method

GrayScaling, ForegroundEnhance, Median softening filter is applied for input image, respectively. Then apply the Canny edge detection method.

Canny Edge Detector

The Canny operator is usually applied in one-dimensional edge detection. As for the step edges, the optimized edge detection pattern obtained with Canny is similar to the first order derivative of the Gaussian function. Using the symmetry and decomposable property of two-dimensional Gaussian function, we can easily calculate the derivative on any orientation and folds of images. Therefore, we can use the near-optimized sensing operator for step-edge in practical applications as in Equation 6.

$$G(x,y,\sigma) = \frac{1}{2\pi\sigma^2} \exp\left(-\frac{1}{2\sigma^2}(x^2+y^2)\right) \qquad (6)$$

where σ is the Gauss filtering parameter. $f(x,y)$ shows the image function. This image is softened as in Equation 7 to obtain $g(x,y)$.

$$g(x,y) = f(x,y) \times G(x,y,\sigma) \qquad (7)$$

The gradient vector form with first order differential coefficients of the softened image is obtained as in Equation 8.

$$\begin{bmatrix} g_x(x,y) \\ g_y(x,y) \end{bmatrix} = f(x,y) \times \begin{bmatrix} G_x(x,y,\sigma) \\ G_y(x,y,\sigma) \end{bmatrix} \qquad (8)$$

In this model, the gradient value is $\sqrt{g_x^2 + g_y^2}$ and the orientation angle is calculated as in Equation 9.

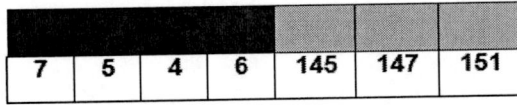

Figure 3. Edge detection.

$$\theta(x,y) = arctan(g_x/g_y) \quad (9)$$

where we get the maximum of the edge points in the gradient angle direction. If we change the σ value of the Gauss function, the Gauss window will change in size. The gradient size of g (x, y) is calculated as in Equation 10.

$$g_{xy}(x,y) = max\left(\sqrt{g_x^2(x,y) + g_y^2(x,y)}\right) \quad (10)$$

Throughout the image, if the gradient size G (x, y) of the point (x, y) is not greater than the two adjacent interpolation in the θ (x, y) direction, the margin point is marked as non-border, otherwise it is marked as the edge point [38]. Double threshold (Low and High) algorithm is applied to detect and connect edges.

In the most basic description, edge detection algorithms are determined by the difference in color values of pixels on the image.

When we look at Figure 3, we can predict where differentiation begins. As you can see, there is a sharp color transition between the matrix elements numbered 6 and 145, which means the starting and ending lines of two different objects. These transitions represent the border lines.

Multilayer Neural Network and Training

Our brain is made up of millions of neurons, so a neural network is a combination of perceptrons that are connected in different ways and that operate in different activation functions.

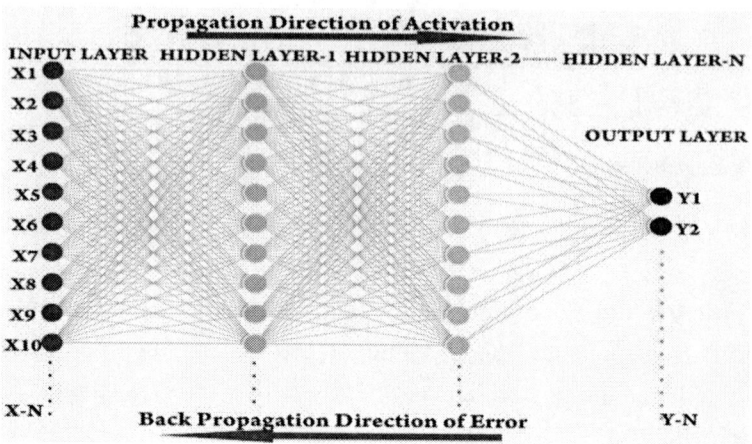

Figure 4. MLP neural network.

The input layer provides information to the network from the outside world. No calculation is performed on any of the input nodes. They only transfer information to the hidden layer. The hidden layer has no direct connection to the outside world. The required calculations are made and the information is transferred from the input layer to the output layer. A network can have only one input layer and a single output layer, but the number of hidden layers can be zero or many. A MLP can have one or more hidden layers. The output layer is responsible for information process and transmitting information to the outside world of the network.

Artificial neural networks, inspired by the biological nervous system, are the computer systems frequently used in the field of machine learning. In these networks, learning takes place through examples [39, 40]. The network architecture is multilayered, as shown in Figure 4.

When creating network architectures, the number of hidden layers is usually determined by trial and error method [41, 42].

This study was performed by learning with back propagation algorithm. In back propagation learning, the network transmits the results it produces on the output according to each input value to hidden layers. The values obtained in order to find the errors that occur at the output are compared with the actual values. The derivative of these errors is calculated and spread to hidden layers in the back. The neurons in the hidden layers use their own weight to reduce their errors. The crucial factor in the structure of the neural network is that; the correct choice of number of neurons in the hidden layer. Selecting a small number of neurons makes learning difficult, and a large number of neurons scatter the ability to generalize. This has a negative effect on the teaching process.

The number of hidden layers is 1. The number of neurons in the hidden layer is equal to the input parameter value because it gives good results in this study. That is, the number of neurons in the hidden layer is a dynamic value that occurs as a result of each segmentation process. The number of neurons in the layers in the formation of the network architecture is also dynamic. Briefly, the network architecture can be defined as follows:

DYNAMIC_INPUT(N)   X   1_COUNT_HIDDENLAYER   X   DYNAMIC_OUT (N).

In the MLP, each neuron with index j in the hidden layer sums the output by multiplying the w_{ij} weight of x_i input. The output of y_j is calculated as a function of the sum according to Equation 11.

$$y_j = f\left(\sum w_{ji} x_i\right) \tag{11}$$

where f is the activation function and is required to convert the weighted sum of the signals in neurons. The activation function is usually sigmoid. If the sigmoid function is preferred, the output value of the j. neuron in the hidden layer is calculated according to Equation 12.

$$\varsigma_j^a = \frac{1}{1+e^{-(NET j^a + \beta j^a)}} \tag{12}$$

The value given as βj^a is the weight of the threshold value element connected to jth. The output of this threshold value element is equal to 1 and this value is constant.

The square of the sum of the differences between the desired and real value in the output neuron is expressed as E in Equation 13.

$$E = \frac{1}{2}\Sigma_j(y_{dj} - y_j)^2 \tag{13}$$

where y_{dj}, j is the desired value of the indexed output neuron and y_j is the real value of this output neuron. Once the MLP neural network calculates the error at the network output, it updates the neuron weights to reduce the error. To rearrange each w_{ji} weight, Δw_{ji} are added to weights. The calculation of Δw_{ji} depends on the training algorithm [41, 42].

Back Propagation Algorithm (BPA)

According to the BPA developed by Rumelhart et al.[43], the change in the link between the i and j indexed neurons in k-th iteration is obtained according to Equation 14:

$$\Delta w_{ji}(k) = -a\frac{\partial E}{\partial w_{ji}(k)} + \mu \Delta w_{ji}(k-1) \tag{14}$$

where α is the learning coefficient, μ is the momentum coefficient, and $\Delta w_{ji}(k-1)$ is the weight difference in the previous iteration. It is effective when the BPA approaches the absolute minimum for its appropriate values.

BPA may need to be trained for a long time to minimize the weight errors. The delta-bar-delta algorithm is a training algorithm developed to accelerate the training of the MLP neural networks.

The Delta-Bar-Delta Algorithm

The delta-bar-delta training algorithm, which adjusts all weights in the network, is optionally used. $\Delta w_{ji}(k)$ the change in connection between the i and j indexed neurons in the k-th iteration is obtained according to Equation 15 [44].

$$\Delta w_{ji}(k) = -a_{ji}(k) \frac{\partial E}{\partial w_{ji}(k)} \tag{15}$$

where $a_{ji}(k)$ is the learning coefficient of the link between i and j indexed neurons.

Dataset Numeric Data

Table 1 shows the number of teeth used in the 4-21 age group. Age and gender information of these tooth images are also demonstrated.

Table 1. 4-21 Age DataSet

Age	Gender	Count	Age	Gender	Count	Age	Gender	Count
4	F	2	9.5	M	5	15.5	F	7
4	M	4	10	F	11	15.5	M	4
4.5	F	3	10	M	7	16	F	5
4.5	M	6	10.5	F	7	16	M	6
5	F	8	10.5	M	10	16.5	F	10
5	M	9	11	F	5	16.5	M	5
5.5	F	12	11	M	11	17	F	5
5.5	M	11	11.5	F	8	17	M	4
6	F	15	11.5	M	4	17.5	F	9
6	M	18	12	F	4	17.5	M	6
6.5	F	17	12	M	2	18	F	9
6.5	M	13	12.5	F	2	18	M	5
7	F	19	12.5	M	4	18.5	F	12
7	M	16	13	F	4	18.5	M	2
7.5	F	14	13	M	5	19	F	10
7.5	M	16	13.5	F	5	19	M	8
8	F	17	13.5	M	6	19.5	F	16
8	M	16	14	F	7	19.5	M	9
8.5	F	9	14	M	4	20	F	3
8.5	M	8	14.5	F	4	20	M	4
9	F	15	14.5	M	5	20.5	F	7
9	M	8	15	F	4	20.5	M	7
9.5	F	6	15	M	4	21	F	9

APPLICATION

Panoramic images are prepared for cropping and testing. In this process, when the image of the tooth is cropped (C), the zoom operation (A) is performed simultaneously and the boundary regions (B) become fully visible.

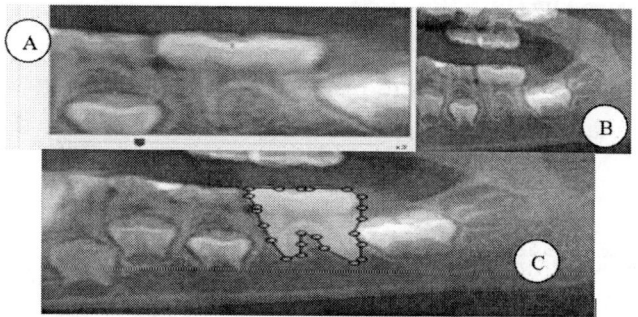

Figure 5. Image editing for test.

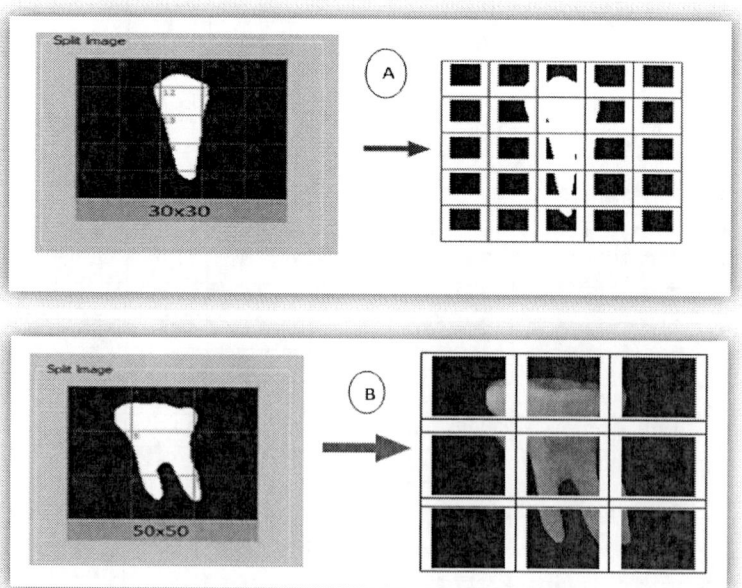

Figure 6. Segmentation process.

In order to present the input parameters of the MLP neural network, images are first passed through certain processes in the application. The image is dynamically divided into sub-segments. Each image divided into sub-segments was applied to a feature extraction with the mean absolute deviation method. This happens if the full segment option is active. So it applies to all sub-segments. If the Full segment option is not active, then the feature reduction is applied.

The numbering to the segments starts from the top left corner and the first column is completed as shown in Figure 6. Then it passes to column 2. Segment 3 shown in Figure 6 (B) is completely black on all teeth. This segment can optionally be used in weight reduction. Optionally, it can also be used for weight reduction in areas that are completely white. The average absolute deviation applied to these segments as a result of the weights (there is a scroll bar) were shown in Figure 7.

Feature Extraction Process

Dental images are divided into sub-images as desired by the user (10 * 20, 50 * 5 etc.). For example, when we divide the 260x160 image into sub-images 20x20 in size, there will be 13 * 8 = 104 sub-images. The feature was extracted by using the mean absolute deviation method. Application code of this operation to sub-images was already shown in Algorithm 1. This process is applied according to Equation 16.

$$F = \frac{1}{N} \left(\sum_{i=1}^{N} |a(x,y) - m| \right) \qquad (16)$$

where N represents the total number of pixels in the image, and m is the mean of all pixel values in the image. $a(x, y)$ is the pixel value at (x, y). In Figure 6. 74, 75. and the 90th segment indicates that it is completely black for all teeth. Figure 7 shows the status of the segments and numerical data resulting from the feature extraction process.

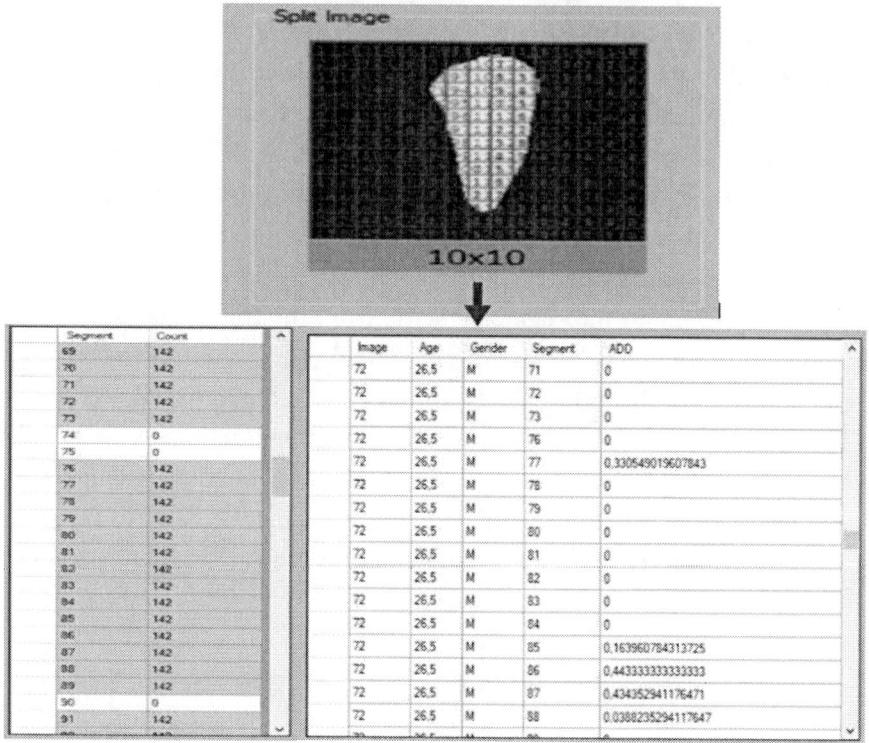

Figure 7. Segmentation process result.

RESULTS

The learning rate as input parameter to the MLP neural network was 0.1 and the sigmoid alpha value was 2. Other input parameters are dynamic and specially created for each training. The teeth were kept in separate folders according to their developmental period (4-9 years, 10-14 years, 15-22 years) and classification was made.

The images of the teeth between the ages of 4-9 the input images are shown in Figure 8.

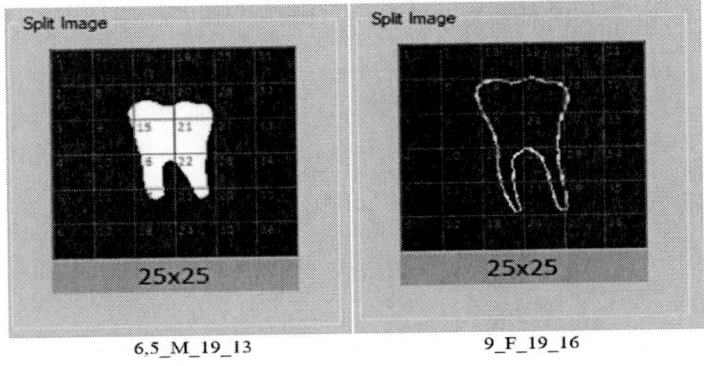

Figure 8. Results of M1-M2 application.

The explanations of the terms used for common areas in the following tables are as follows: M: Method, IS: Image Size, NS: Number of Segments, IN: Iteration Number, NNHL: Number of Neurons in the Hidden Layer, NT: Number of Training, TN: Test Number, AE: Averagre Error, IC: Input Count, RA: Real Age, RG: Real Gender, EA: Estimated Age, EG: Estimated Gender, D: Difference, CS: Classification success.

The input parameter information of the teeth between the ages of 4-9 for network is shown in Table 2.

The results obtained without any reduction (full segment) in the segments of the teeth between the ages of 4-9 are shown in Table 3.

Table 2. Input parameter values

M	IS	NS	IN	NNHL	NT	TN	AE	IC
M1	150x150	25X25	52342	24	263	6	0.16	36x257
M2	150x150	25X25	8453	24	261	8	0.16	36x257

Table 3. Full segment results

M	RA	RG	EA	EG	D	CS
M1	6	M	6.5	M	+-0.5	%91
M1	8	M	8	M	+-0	%99
M2	6,5	M	5,5	M	+-1	%32
M2	9	F	8,5	M	+-0.5	%96

Table 4. Not Full segment results

M	RA	RG	EA	EG	D	CS
M1	4.5	M	6,5	M	+-2.5	%100
M1	5	M	8,5	M	+-3	%100
M1	9	M	8	M	+-1	%100
M2	6	M	8,5	F	+-2.5	%99
M2	9,5	F	9,5	M	+-0	%99

Table 5. Input parameter values

	IS	NS	IN	NNHL	NT	TN	AE	IC
M1	150x150	5X10	14701	20	109	6	9,49	450X109
M2	150x150	30X15	8453	20	109	6	0,09	50X109

Table 6. Full segment results

M	RA	RG	EA	EG	D	CS
M1	10	F	10	F	+-0	%99
M1	14	M	12	M	+-2	%99
M2	10,5	F	13	F	+-2,5	%93
M2	10	F	10	F	+-0	%64

Table 7. Not Full segment results

M	RA	RG	EA	EG	D	CS
M1	11	M	13,5	M	+-2,5	%99
M1	14	M	10	F	+-4	%99
M2	10,5	M	10	M	+-0,5	%99
M2	10	F	10	M	+-0	%99
M2	14	F	13	F	+-1	%99

The results obtained by not reducing the segments of the teeth between the ages of 4-9 (not full segment) are shown in Table 4.

The input parameter information of the teeth between the ages of 10-14 for network is shown in Table 5.

The results obtained without any reduction (full segment) in the segments of the teeth between the ages of 10-14 are shown in Table 6.

Table 8. Input parameter values

M	GB	SS	İS	AKNS	ES	TS	OH	GS
M1	150x150	50X25	2051	32	228	9	0,15	18X228
M2	150x150	30X15	2670	32	227	10	0,15	50X227

Table 9. Full segment results

M	RA	RG	EA	EG	D	CS
M1	19	M	19	M	+-0	%99
M1	20,5	M	19	M	+-1,5	%98
M1	21	M	20	M	+-1	%99
M2	16,5	F	17,5	F	+-1	%76

Table 10. Not Full segment results

M	RA	RG	EA	EG	D	CS
M1	19.5	F	18	M	+-1.5	%99
M1	20,5	M	19	M	+-1,5	%99
M1	21	M	18	M	+-3	%99
M2	17,5	F	19,5	F	+-2	%99

The results obtained by not reducing the segments of the teeth between the ages of 10-14 (not full segment) are shown in Table 7.

The input parameter information of the teeth between the ages of 15-22 for network is shown in Table 8.

The results obtained without any reduction (full segment) in the segments of the teeth between the ages of 15-22 are shown in Table 9.

The results obtained by not reducing the segments of the teeth between the ages of 15-22 (not full segment) are shown in Table 10.

When the above tables are examined, weight and gender estimations are more accurate in the weight reduction process (full segment). However, classification success was lower. In the weight reduction process, the classification rates were higher. As cumulative success is more important, it can be said that the weight reduction process yields better classification accuracy. That is, better results can be obtained when the test images change.

CONCLUSION

In this study, artificial intelligence and image processing techniques were used. In some cases, the age and gender determination process determined by the forensic sciences was performed automatically in the computer environment. The boundaries of the dental X-ray images were removed to create the database. The database was created manually with 562 images. There are a total of 69 different age groups in this database consisting of F and M groups. The teeth were grouped according to the stages of development (4-9 age, 10-14 age, 15-22 age). Some pre-processing techniques were performed on these teeth. After pre-processing, M1 and M2 methods were converted to binary form. With the MLP neural network, some of these groups were used in the network training and some of them were used in the testing process of the network. The images were first extracted from their attributes after they were segmented in a dynamic structure, and finally the vertebrates were created. The obtained feature vectors were classified by MLP neural network with the highest % 100 classification accuracy. The gender estimation process was also performed.

REFERENCES

[1] Yılmazer, Ö. (2006). *Adli Tıp Kurumu'nda yaş tayininde kullanılan yöntemin verimlilik açısından değerlendirilmesi* [Evaluation of the method used in age determination in the Forensic Medicine Institute in terms of efficiency]. Specialist thesis, Istanbul.

[2] Demirkıran, DS., Çelikel, A., Zeren, C., Arslan, MM. (2014) Yaş tespitinde kullanılan yöntemler [Methods used in age determination]. *Dicle Medical Journal*, 41(1):238-43.

[3] Erbudak, HÖ., Ozbek, M., Uysal, S., Karabulut, E. (2012). Application of Kvaal et al.'s age estimation method to panoramic radiographs from Turkish individuals. *Forensic Sci In,,;* 219(1-3):141-6.

[4] Avon, SL. (2004). Forensic odontology: the roles and responsibilities of the dentist. *J Can Dent Assoc.* Jul-Aug, 70(7): 453-8.

[5] Shamim, T., Varghese, V., Shameena, PM., Sudha, S. (2006). Ageestimation: a dental approach. *J Punjab Acad Forensic Med Toxicol*, 6:14-6.

[6] Isır, AB., (2011). *Adli hekimlikte yaş tayini. İçinde: Birinci Basamakta Adli Tıp.* Ed: Koç S, Can M. 2.edt., Istanbul: Golden Print: pp. 222-35. ISBN: 978- 605-5867-33-1.

[7] Patil, SK., Mohankumar, KP., Donoghue, M. (2014). Estimation of age by Kvaal's technique in sample Indian population to establish the need for local Indianbased formulae. *J Forensic Dent Sci.,* 6(3):166-70.

[8] Someda, H., Saka, H., Matsunaga, S., Ide, Y., Nakahara, K., Hirata, S., Hashimoto, M. (2009). Age estimation based on three-dimensional measurement of mandibular central incisors in Japanese. *Forensic Sci Int.,* 185(1-3):110-4.

[9] Karkhanis, S., Mack, P., Franklin, D. (2014). Age estimation standards for a Western Australian population using the dental age estimation technique developed by Kvaal et al. *Forensic Sci Int.,* 235:104.

[10] Karadayı, B. (2010). *Dişlerden erişkin ve erişkin olmayan bireylerden yaş belirlenmesi: dijital radyolojik teknik uygulamaları.* Institute of Forensic Medicine, Institute of Science, PhD Thesis, 123-6.

[11] Rai B, Kaur J. *Evidence-Based Forensic Dentistry.* Springer (2013). ISBN: 978-3-642-28993-4.

[12] Birdal, R. G., Gumus, E., Sertbas, A., Birdal, İ. S. (2016). Automated lesion detection in panoramic dental radiographs, *Oral Radiol, Japan*, 32:111–118.

[13] Dinçer, İ., (2015). *Adli tıpta yaş tayininde dişlerin muayenesi ile elde edilen bilgilerin değerlendirilmesi* [Evaluation of information obtained by examination of teeth in age determination in forensic medicine], Graduation thesis, Ege University, Faculty of Medicine, Department of Forensic Medicine, İzmir.

[14] Eyad Haj Said, D. E. (2006). Teeth Segmentation in Digitized Dental X-Ray Films Using Mathematical Morphology, *IEEE transactions on information forensics and security.*

[15] Eyad Haj Said A. A. (2008). Accurate Segmentation Of Digitized Dental X-Ray Records, *IEEE.*

[16] Eyad Haj Said, G. F. (2004). Dental X-ray Image Segmentation, Biometric Technology for Human Identification, *Proceedings of SPIE.*

[17] Lin, PL., Huang, PY., Huang, PW., Hsu, HC., Chen, CC. (2014). Teeth segmentation of dental periapical radiographs based on local singularity analysis, *Comput Methods Progr Biomed*, 113:433–45.

[18] Mourıtsen, D. A. (2013). *Automatıc Segmentatıon of Teeth in Dıgıtal Dental Models*, Master Thesis, The University of Alabama at Birmingham, UK.

[19] Nimbalkar, S. (2016). *Accuracy of Volumetric Analysis Software Packages in Assessment of Tooth Volume Using CBCT*, School of Dentistry in conjunction with the Faculty of Graduate Studies, Master Thesis, Loma Lında Unıversity, ABD.

[20] Rad, A. E., Rahim, M. S. M., Kumoi, R., Norouzi, A. (2012). Dental x-ray image segmentation and multiple feature extraction, *2nd World Conference on Innovation and Computer Sciences,* 2: 188-197.

[21] Al-sherif, N. (2013). *Novel Techniques for Automated Dental Identification*, PhD Thesis, West Virginia University, ABD.

[22] Yun, JI, Lee, JY, Chung, JW, Kho, HS, Kim, YK. (2007). Age estimation of Korean adults by occlusal tooth wear. *J. Forensic Sci.,* 52 (3):678-83.

[23] Blenkin, MRB. (2005). *Forensic Dentistry and its Application in Age Estimation from the Teeth using Modified Demirjian System*, Master Thesis, Sidney Üniversitesi, Avusturalya.

[24] Cruz-Landeira, A., Linares-Argote, J., Martínez-Rodríguez, M., Rodríguez-Calvo, MS., Otero, XL., Concheiro, L.(2009). Dental age estimation in Spanish and Venezuelan children. Comparison of Demirjian and Chaillet"s scores, *Int J Legal Med.,* 124 (2):105.

[25] Cameriere, R., Ferrante, L. (2008). Age estimation in children by measurement of carpals and epiphyses of radius and ulna and open apices in teeth: a pilot study, *Forensic Sci. Int.* 174:60–63.

[26] Cameriere, R., Giuliodori, A., Zampi, M., Galic, I., Cingolani, M., Pagliara, F. et al. (2015). Age estimation in children and young adolescents for forensic purposes using fourth cervical vertebra (C4), *Int. J. Legal Med.,* 129:347–355.

[27] Nystrom, M., Peck, L., Kleemola-Kujala, E., Evalahti, M., Kataja, M. (2000). Age estimation in small children: reference values based on counts of deciduous teeth in Finns, *Forensic Sci. Int.,* 110: 179–188.

[28] Cameriere, R., Angelis, De, D., Ferrante, L., Scarpino, F., Cingolani M. (2007). Age estimation in children by measurement of open apices in teeth: a European formula, *Int. J. Legal Med.,* 121:449–453.

[29] Paewinsky, E., Pfeier, H., Brinkmann, B., (2005). Quantification of secondary dentine formation from orthopantomograms–a contribution to forensic age estimation methods in adults, *Int. J. Legal Med.,* 119: 27–30.

[30] Guo, Y. C., Yan, C. X., Lin, X. W., Zhou, H., Li, J. P., Pan, F. et al. (2015). Age estimation in northern Chinese children by measurement of open apices in tooth roots, *Int. J. Legal Med.,* 129: 179–186.

[31] Schmidt, S., Baumann, U., Schulz, R., Reisinger, W., Schmeling, A. (2008). Study of age dependence of epiphyseal ossification of the hand skeleton, *Int. J. Legal Med,.* 122: 51–54.

[32] Schmidt, S., Nitz, I., Schulz, R., Schmeling, A. (2008). Applicability of the skeletal age determination method of Tanner and Whitehouse for forensic age diagnostics, *Int. J. Legal Med.,* 122:309 –314.

[33] Garvin, H. M., Passalacqua, N. V., Uhl, N. M., Gipson, D. R., Overbury R. S., Cabo L. L. (2012). Developments in forensic anthropology: age-at-death estimation, *A Companion to Forensic Anthropology*, John Wiley Sons, Ltd., 202–223.

[34] Avuçlu, E., Başçiftçi, F. (2019). Novel Approaches To Determine Age And Gender From Dental X-Ray Images By Using Multiplayer Perceptron Neural Networks And Image Processing Techniques, *Chaos Solitons Fractals*, 120: 127-13.

[35] Avuçlu E., Başçiftçi, F. (2018). New Approaches to determine Age and Gender in Image Processing Techniques using Multilayer Perceptron Neural Network, *Applied Soft Computing*, 70: 157–168.

[36] Avuçlu, E., Başçiftçi, F. (2018). Determination age and gender with developed a novel algorithm in image processing techniques by implementing to dental X-ray images, *Romanian Journal of Legal Medicine*, 26(4):412-418.

[37] Avuçlu E. (2019). *Chronicological Age Determination From Dental X-Ray Images by Using Artificial Intelligence And Image Processing Techniques*, PhD. Thesis, The Graduate School of Natural And Applied Science of Selçuk University Computer Engineering.

[38] Chen, Y., Deng, C., Chen, X. (2015). An Improved Canny Edge Detection Algorithm. *International Journal of Hybrid Information Technology*, 8(10):359-370.

[39] Tafeit, E., Reibnegger, G. (1999). Artificial Neural Networks in Laboratory Medicine and Medical Outcome Prediction, *Clinical Chemistry and Laboratory Medicine*, 37(9): 845-85.

[40] Güler, İ., Übeyli, E. D. (2003). Detection of Ophthalmic Artery Stenosis by Least-Mean Squares Backpropagation Neural Network, *Computers in Biology and Medicine,* Vol 33, No 4, 333-343.

[41] Haykin S. (1994). *Neural Networks: A Comprehensive Foundation*, New York, Macmillan College Publishing Company Inc.

[42] Chaudhuri, B. B., Bhattacharya, U. (2000). Efficient Training and Improved Performance of Multilayer Perceptron in Pattern Classification, *Neurocomputing*, 34: 11-27.

[43] Rumelhart, D. E., Hinton, G. E., Williams, R. J. (1986). Learning Representations by Back-Propagating Errors, *Nature*, 323:533-536.

[44] Jacobs, R. A. (1988). Increased Rate of Convergence through Learning Rate Adaptation, *Neural Networks*, 1: 295-307.

In: Multilayer Perceptrons
Editor: Ruth Vang-Mata

ISBN: 978-1-53617-364-2
© 2020 Nova Science Publishers, Inc.

Chapter 4

DYNAMIC FORECASTING OF ELECTRIC LOAD CONSUMPTION USING ADAPTIVE MULTILAYER PERCEPTRON (AMLP)

Jeremias T. Lalis[*] *and Elmer A. Maravillas*

College of Computer Studies, Cebu Institute of Technology - University Cebu City, Cebu, Philippines

ABSTRACT

Electric energy plays a vital role in the achievement of social economic and environment development of any nation. Thus, efficient demand planning and production of energy is needed to avoid too much over/under-estimation of electric load. In this study, the researchers proposed a scheme with eight steps for a dynamic time series forecasting using adaptive multilayer perceptron with minimal complexity. Two different data sets; each divided into three overlapping parts (training, validating and testing sets), from two different countries were used in the experiments to measure the robustness and accuracy of the models

---

[*] Corresponding Author's Email: jeremias.lalis@gmail.com.

produced by the AMLP. Experiments results show the effectiveness of the proposed scheme for AMLP in forecasting the electric load consumption based on the calculated coefficient of variance of RMSD, CV (RMSD).

Keywords: electric load consumption, long-term forecasting, backpropagation, adaptive multilayer perceptron

INTRODUCTION

Electric energy plays a vital role in the achievement of social, economic and environmental development of any nation. It is a significant driving force for the modernization, growth in the economy and improvement of quality of life. The rising number of population and standard of living caused the demand for energy to increase. Moreover, it has been shown in the annual report published by UN, the Human Development Index report, that a small increase in the consumption of electricity implies a significant growth in the gross domestic product (GDP). This is probably due to industry development and increase of production of both, developed and developing countries.

However, the rapid increase of raw-material prices due to resource exhaustion, the need to reduce greenhouse gases emission and avoiding unnecessary excessive costs have caused the power industries not to over-produce electrical energy. On the other hand, under-estimation of load causes insufficient reserves of power, discouraging any industrial and economic developments. Thus, efficient demand planning and production of energy require immediate attention. It is necessary to develop a dynamic long-term power load forecasting systems in order to produce accurate and reliable prediction models and to actively cope with the changes in the power industry.

And because of these reasons, electric load consumption forecasting problem is now receiving great and growing attention. Various studies with varying statistical and technical methods; such as autoregressive integrated moving average (ARIMA) (Chujai, Kerdprasop, et al., 2013), multiple linear regression (MLR) (Badran and Abouelatta, 2013) and artificial neural networks (ANN) (Kandananond, 2011) (Arroyo, Skov, and Huynh, 2005) (Yalcintas and Akkurt, 2005) (Simsar, Alborzi, et al., 2013), had already been conducted in the last decades to forecast energy consumption. In the study of (Kandananond, 2011) (R. E. Edwards, et al., 2012), it has been shown that the ANN outperformed the other methods in producing accurate prediction models due to its capability to learn by approximating the linear or nonlinear relationship between the input and output values. However, ANN is still facing several challenges. Yang, Rivard, and Zmeureanu, 2005 observed that most of the surveyed literatures involved a single prediction model, wherein, existing ANN models are trained using huge amount of historical data to obtain static model(s) but failed to update its parameters as new set of significant data became available. Furthermore, R. E. Edwards, et al., 2012 cited two major problems of gradient descent learning method; over-fitting and avoiding local minima. And based on the experiments conducted by (Arroyo, Skov, and Huynh, 2005), the determination of the number of epochs and learning rate are necessary to obtain optimal prediction models but are equally difficult to do. And according to (Boniface, Alexandre, and Vialle, 1999), training the ANN with a large volume of data will require longer time, relative to its computational cost, in order for it to learn.

DYNAMIC FORECASTING

Electric load forecasting has been an attractive research topic in different countries all over the world for the past years. It is because of the fact that electric energy is one of the key factors for the sustainable development of social, economical and environmental aspects of any nation. In the surveyed literatures, load forecast is generally classified into four categories based on forecasting time: very short-term, short-term, medium-term, and long-term forecasting.

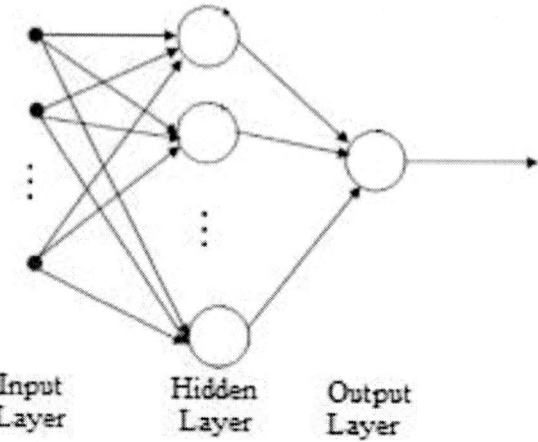

Figure 1. The multilayer perceptron architecture.

In this study, the researchers are focusing on the development of an adaptive multilayer perceptron for an accurate long-term forecasting of electric load consumption.

A. Long-Term Forecasting

Badran and Abouelatta, 2013 hybridized the traditional artificial neural network by integrating the multiple linear regression method on it to obtain the predicting load demand for up to year 2020. Results show the feasibility of this method in time-series prediction. Kandananond, 2011, compared the performance of the three different forecasting approaches - multiple linear regression (MLR), autoregressive integrated moving average (ARIMA) and artificial neural networks (ANN) - by utilizing the historical data of Thailand from 1986 to 2010. The historical data is composed of country's population, gross domestic product and electricity consumption. The study shows that ANN approach outperformed the other two methodologies based on the data set used. Kazemi et al., 2011 estimated the annual residential and commercial energy demand of Iran for the period of 2008 to 2020 using their prediction model. The model was created using neuro-based approach. Several socio-economic indicators -

such as total number of households, energy consumption, energy prices, construction investment and gross domestic product - were considered in building the model. Model predictions were compared with the evaluation stage data. Results show the validity of the model in producing predictions of electric load demand.

B. Multilayer Perceptron

The multilayer perceptron (MLP) is a fully connected network model that maps the input data sets into the corresponding output sets. It was the first and simplest type of neural network composed by multiple layers of nodes known as the input layer, hidden layer, and the output layer. Each layer is connected to the next layer through its processing elements known as neurons in weighted links. Each neuron has nonlinear activation function that scales its net input into a specific range. In the case of multiple hidden layers, the output from one hidden layer is forwarded into the following hidden layer and separate weights are provided to the sum going into each layer.

C. The Standard Backpropagation Algorithm

In training the network, MLP learns through a supervised learning technique called backpropagation (BP) learning algorithm. The standard BP (Duda, Hart, et al. 2001) for forecasting works as follows:

1. Initialize network weights W to small pseudorandom values.
2. Calculate the activation level O_j of the hidden and output units using a sigmoid function F,

$$F(v) = tanh(v) \qquad (1)$$

3. Calculate the error using the delta rule,

$$E = \Sigma j 1/2(t_j - O_j)^2 \qquad (2)$$

where t_j is the predicted value and O_j is the actual value at the output layer.

4. Update the network weights by computing W_{ji} for all weights from output to hidden layer,

$$\Delta W_{ji} = \eta \delta_j O_i \qquad (3)$$

5. Repeat step 2 to 5 until the stopping criterion is being satisfied.

D. Serial Correlation in Time Series

MLP architecture consists of three layers, the input, hidden and output layers. The study of (Lin, Yu, et al. 1995) shows how the number of nodes in the input layer affects the forecasting performance of the neural network. Carefully choosing the number of inputs will lead to a better prediction outcome. Serial correlation analysis is one of the effective ways to determine the number of inputs for the model in time series forecasting. It is also being used in different statistical approaches such as ARIMA models. In this study, the researchers applied serial correlation analysis to automate the determination of the number of input nodes based on the presented time window. The autocorrelation in lag r_k, the correlation between x_i and x_{i+k}, can be computed by

$$r_k = \frac{\sum_{i=1}^{n-k}\frac{(x_i - \bar{x})(x_{i+k} - \bar{x})}{n-k}}{\sum_{i=1}^{n}\frac{(x_i - \bar{x})^2}{n}} \qquad (4)$$

E. The Adaptive Weights (Nguyen-Widrow Randomization)

MLP needs to be trained with sufficient volume of data and in different architecture in order for it to learn and produce a good prediction model. However, these will require longer time and processing power. In order to minimize the training time, the researchers adapted the Nguyen-Widrow randomization technique (Nguyen and Widrow, 1990) in initializing the weight of each neuron in the hidden layer. This technique has sufficiently increased the trainability of each network, thus, speeding up the training process. To implement Nguyen-Widrow:

1. Initialize neural network weights W with small pseudorandom values.
2. Calculate the Beta β using the number of hidden h and input i neurons,

$$\beta = 0.7h^{1/I} \tag{5}$$

3. Get the Euclidean norm n of all weights W in the hidden layer using the following formula,

$$n = sqrt\left(\sum_{i=0}^{i<wmax}(Wi)^2 \right) \tag{6}$$

4. Adjust the weight value of each neuron in the hidden layer using the previously computed beta β and norm n values, as shown below,

$$w_{t+1} = \beta w_{t+1}/n \tag{7}$$

F. The Adaptive Stopping Criterion

According to (Arroyo, Skov, and Huynh, 2005), to obtain optimal prediction models it is necessary to determine the number of epochs and learning rate. However, these are difficult to do especially in the case of sliding window training since the MLP will be trained with multiple patterns in different time windows. The researchers used an adaptive stopping criterion for BP algorithm, as proposed by (Lalis, Gerardo, and Byun, 2014), to address these problems and to enable the MLP to be trained in a real-time manner. Furthermore, optimal prediction models were also obtained since the stopping criterion enabled the networks to converge on the global minimum based on the presented patterns on it. The adaptive stopping criterion algorithm works as follows:

1. Set the value of threshold and counter to zero (0), the previous Median Squared Error (MdSE) and least MdSE to one (1).
2. Initialize the predefined network.
3. Calculate the MdSE of the network based on the presented patterns (epoch) using the standard BP learning method.
4. Round off the MdSE to the nearest five decimal places.
5. If the previous MdSE is equal to the MdSE decrement counter by one (1), otherwise, increment counter by one (1).
6. Override the value of the previous MdSE with the calculated MdSE.
7. Assign the value of MdSE to the least MdSE if it is less than the least MdSE.
8. Check if the value of counter is equal to zero (0). If it is equal to zero (0):
   - Calculate the value of the threshold with the following formula,

$$\text{Threshold} = (\text{threshold} * 3 + \text{least MdSE})/2 \qquad (8)$$

- Assign the calculated threshold to least MdSE if it is greater than the least MdSE.
- If the MdSE is less than the threshold, stop the training, otherwise, start again from step two (2) until the network stops learning.
9. If counter is not equal to zero (0), start again from step three (3) until the network stops learning.

Minimal changes in the movement of the patterns based on the calculated delta is signified by rounding off the median squared error into five decimal places.

G. Sliding-Window Training

Adaptive multilayer perceptron (AMLP) models can be trained in real-time and online manner due to their capability to adapt themselves in the unprecedented changes on the patterns from incoming data. As new operational and environmental data become available, the AMLP can be retrained to produce better prediction model. There are two ways to train AMLP, first is the accumulative training technique and the other one is the sliding window training technique. This study is focusing on the use of sliding window training technique as it shows better outcome (Yang, Rivard, and Zmeureanu, 2005) compared to the other training technique.

In sliding window training strategy, the size of the training set is kept constant. But as time goes by and as new measurements and/or significant data become available, the oldest data are removed from the training set and replaced by the new one.

H. The Adaptive Multilayer Perceptron Algorithm

This study proposes a scheme for a dynamic time series forecasting with multilayer perceptron (MLP).

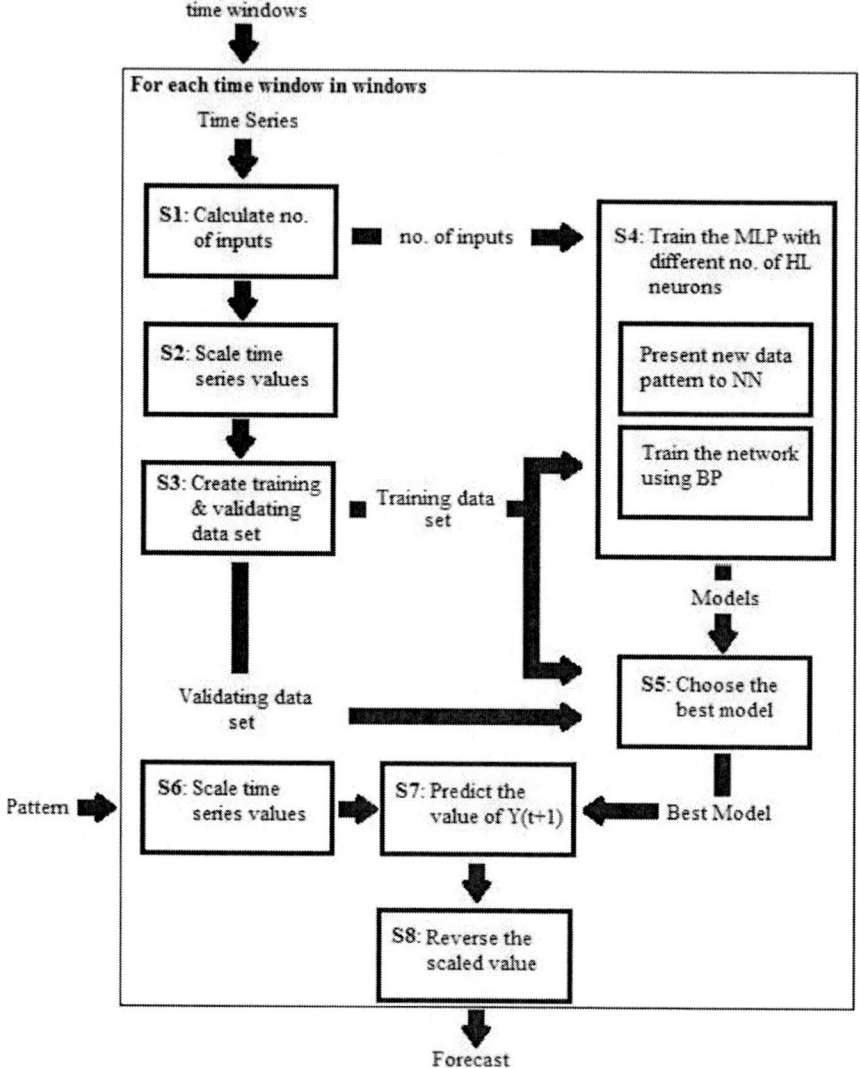

Figure 2. Block diagram of adaptive multilayer perceptron.

The scheme includes eight steps as shown in the Figure 2.

1. Determine the number of inputs by computing the autocorrelation coefficients of the given time series (window). Then, compare the lag of r_k (where k = 0) with $r_{k+1}, r_{k+2}, \ldots, r_n$, the index of the lag that

has largest difference with r_k will be the number of input nodes in the network.

2. Scale the time series values,

$$Y_t = X_t/c \qquad (9)$$

where c is 1 concatenated with 0s. The number of zeros is equivalent to the number of digits of the maximum value in the time series.

3. Create training and validating data sets based on the calculated number of input layer nodes i, where,

$$training\ size = window\ size - i\ div\ 2 - i + 1 \qquad (10)$$

$$validating\ size = window\ size - training\ size - 1 \qquad (11)$$

4. Train the network three times for each HL_n, such that

$$n = i\ div\ 2 + 1,\ i\ div\ 2 + 2,\ \ldots,\ i\ div\ 2 + i$$

note that Nguyen-Widrow randomization technique was adapted to speed-up the training phase.

5. Choose the model with the least weighted value W_m using the following equation:

$$W_m = MdAPE(NN[i,\ HL_n,\ trainingDataSet])/3 + 2 * MdAPE(NN[i,\ HL_n,\ validatingDataSet])/3 \qquad (12)$$

where MdAPE is the Median Absolute Percentage Error of the current model.

6. Accept and scale the pattern for prediction.

7. Forecast the next time series value Ft using the chosen best model and the scaled pattern.
8. Reverse the scaled forecasted value by

$$Ft = Ft * c \qquad (13)$$

EXPERIMENTS AND RESULTS

To test the robustness and to measure the accuracy of the proposed scheme, two different data sets with different time horizons from two different countries were used in the six experiments. The first set of time series was the electric consumption (in Billion Kilowatt-hours) of the 50 states in USA and the District of Columbia from year 1984 to 2013. This data was taken from the website of US Energy Information Administration. The next set of data is the electric power consumption (kWh per capita) of Canada from year 1984 to 2011 from the World Bank.

Each data set was divided into three overlapping parts: training, validating and testing sets, for every time window. In the subsequent experiments, three window sizes were being used: 10, 15 and 20 years, in a sliding-window manner. For example in the window of 10 years, measurements of the electric consumption associated with the first 10 years were selected as the initial training data set. Once the initial training is done and the prediction for electric consumption for 11th year is generated, the electric consumption corresponding to the first year of the time series is discarded and the actual consumption in the 11th year is added in the window and retraining is carried out.

A. Experiment 1

In the first experiment, electric consumption of USA from 1994 to 2013 was used to train, validate and test the performance of the proposed scheme for AMLP. Figure 3 shows the actual and predicted values of AMLP for 10 years.

B. Experiment 2

In the second experiment, a window size of 15-year was used to train the AMLP and predict electric consumption of USA from year 2004 to 2013. Data from 1989 to 2012 was used to build the models and produce forecast as shown in Figure 4.

C. Experiment 3

This experiment was conducted to test the accuracy of the AMLP using the large data set (1984-2013). Figure 5 shows the result of the experiment.

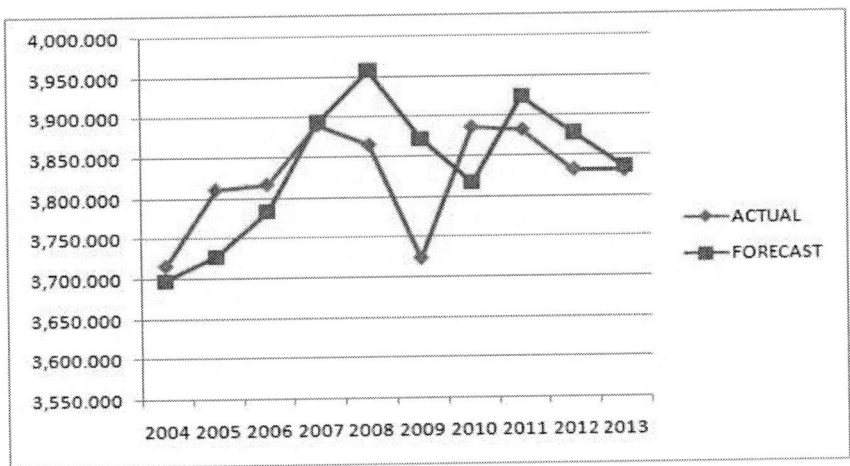

Figure 3. Actual and forecasted electric consumption of USA using 10-year window size.

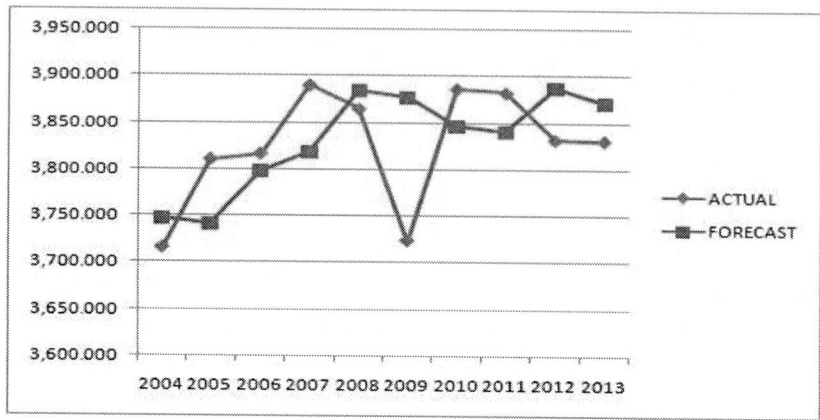

Figure 4. Actual and forecasted electric consumption of USA using 15-year window size.

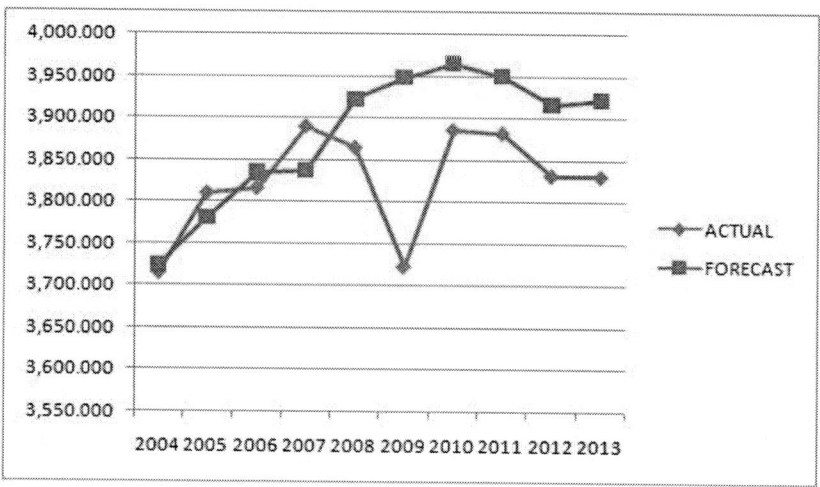

Figire 5. Actual and forecasted electric consumption of USA using 20-year window size.

D. Experiment 4

To test the robustness of AMLP, electric consumption of Canada from 1992 to 2010 was used to train and validate the produced models. The graph in Figure 6 shows the resulting forecast from 2002 to 2011.

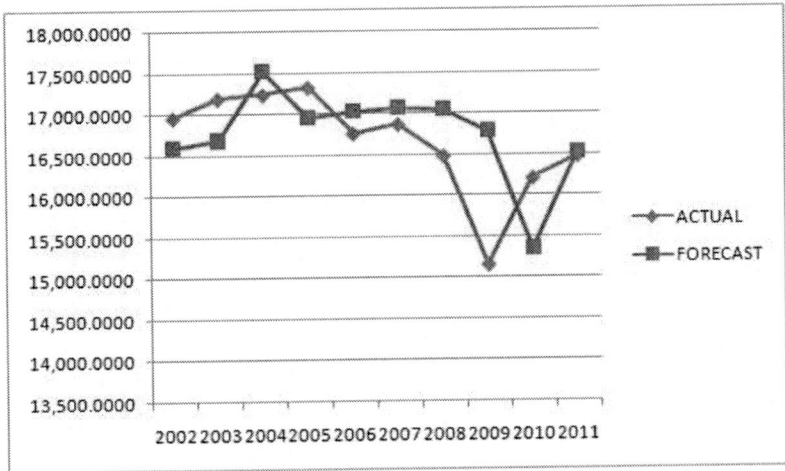

Figure 6. Actual and forecasted electric consumption of Canada using 10-year window size.

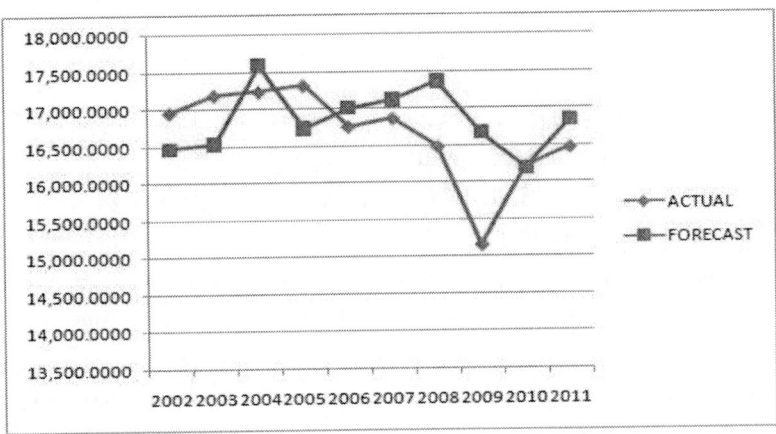

Figure 7. Actual and forecasted electric consumption of Canada using 15-year window size.

E. Experiment 5

In the fifth experiment, electric consumption of Canada from 1987 to 2011 was used to train, validate and test the performance of the proposed scheme for AMLP. Figure 7 shows the actual and predicted values of AMLP for 15 years.

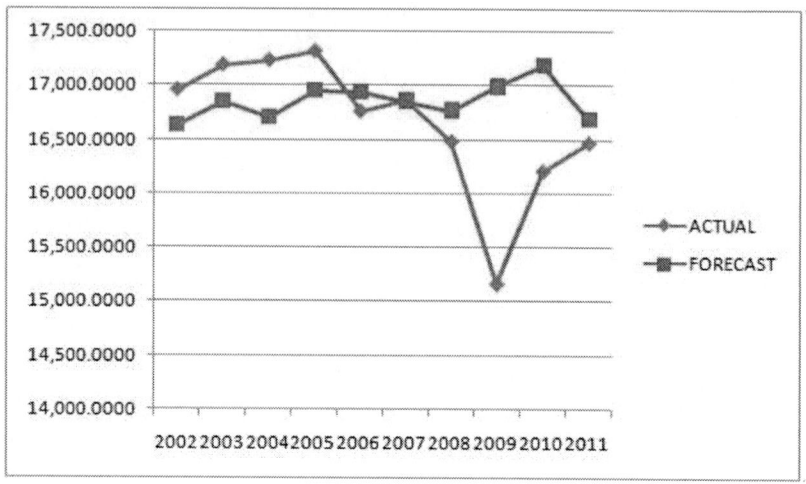

Figure 8. Actual and forecasted electric consumption of Canada using 20-year window size.

F. Experiment 6

This last experiment was conducted to verify the effect of large data set in the performance of the AMLP. Electric consumption of Canada from year 1982 to 2010 was used in this experiment and Figure 8 shows the graph based on the presented testing set. The tables below show the adaptability of the proposed scheme based on the given windows.

Three notable things are being observed based on the above tables. First is that the number of nodes in the input layer is not remaining constant as it move forward to different windows. Second is the optimal number of neurons in the hidden layer is not deterministic, making static ANN prediction models less accurate as time goes by. And finally, using the number of epochs or iteration as stopping criterion for the training phase is less effective in producing optimal prediction model since it is highly dependent on the data presented on the network and neurons' initial weights.

Table 1. AMLP with 10-year window

				10-YEAR					
	Window	ILN#	HLN#	Epoch		Window	ILN#	HLN#	Epoch
U S A	1	4	3	436	C A N A D A	1	4	4	1,968
	2	4	3	2,066		2	3	4	1,110
	3	4	4	1,386		3	4	6	568
	4	4	5	450		4	3	3	2,312
	5	4	4	1,680		5	4	5	840
	6	4	3	1,964		6	4	4	2,112
	7	4	4	818		7	4	5	766
	8	4	4	544		8	4	5	482
	9	4	3	1,364		9	4	3	2,108
	10	4	6	720		10	3	2	2,594

Table 2. AMLP with 15-year window

				15-YEAR					
	Window	ILN#	HLN#	Epoch		Window	ILN#	HLN#	Epoch
U S A	1	6	8	1,828	C A N A D A	1	6	6	798
	2	6	9	1,096		2	6	5	590
	3	6	4	3,734		3	6	9	472
	4	6	4	9,614		4	6	9	364
	5	6	5	1,152		5	6	8	408
	6	6	7	1,850		6	6	7	606
	7	6	9	5,186		7	5	7	820
	8	6	7	744		8	6	9	300
	9	6	5	2,940		9	5	6	748
	10	5	4	440		10	6	5	1,704

The prediction accuracy of the model on each experiment was measured by the root mean squared deviation (RMSD) and coefficient of variation of RMSD, CV (RMSD), denoted as,

$$RMSD = \sqrt{\frac{\sum_{t=1}^{n}[y_{pred}(t) - y_{data}(t)]^2}{n}} \qquad (14)$$

$$CV(RMSD) = \frac{\sqrt{\sum_{i=1}^{n}[y_{pred}(t) - y_{data}(t)]^2}}{|y_{data}|} \times 100 \qquad (15)$$

where: n = the number of data,

y data = the mean of the measured data

ypred(t) = the predicted energy at time t,

ydata(t) = the measured data at time

Table 3. AMLP with 20-year window

					20-YEAR				
	Window	ILN#	HLN#	Epoch		Window	ILN#	HLN#	Epoch
U S A	1	9	9	2,478	C A N A D A	1	8	6	176
	2	8	9	12,948		2	9	13	726
	3	8	8	6,432		3	8	9	1,728
	4	7	5	854		4	9	9	1,518
	5	8	8	338		5	9	11	590
	6	9	13	154		6	9	10	900
	7	9	6	3,112		7	8	11	1,678
	8	9	9	970		8	9	5	766
	9	9	7	1,824		9	9	5	2,372
	10	9	10	380		10	9	11	644

Table 4. Statistical indices for each experiment

Origin	Exp#	Window Size	RMSD	CV(RMSD)	Ave Epoch
USA	1	10 years	68.88	1.80%	1,143
	2	15 years	65.76	1.72%	2,858
	3	20 years	91.86	2.40%	2,949
CANADA	4	10 years	662.99	3.98%	1,486
	5	15 years	665.17	3.99%	681
	6	20 years	711.95	4.27%	1,110

The CV (RMSD) describes how fit is the model in terms of outcome and squared residual values. Low CV signifies small residuals to the predicted value and thus suggestive of a good model fit. Table 4 summarizes the results of all the experiments.

It has been observed from the table above that there is no significant correlation between the size of the window and the number of epochs. This is probably due to the initial value of weights in the networks and the kind of data within the windows.

The table above, using the testing set, shows the stability of the proposed scheme in terms of forecasted value with different origin, time horizon and window sizes. It also shows that the adaptive multilayer perceptron (AMLP) was able to produce an acceptable forecast values based on the calculated CV (RMSD). This signifies the effectiveness of combining the adaptive methods of determining the number of nodes in the input and hidden layers, initialization of weights, and stopping criterion.

Experiments also show that a window size of 15 years provides a reasonable balance between computational complexity and accuracy.

In this paper, the researchers proposed a scheme for an adaptive multilayer perceptron (AMLP) that can produce optimal prediction model with minimal complexity. Electric consumptions of two different countries were divided into three overlapping parts: training, validating and testing set. The training and validating sets were used to train the AMLP and to choose the best model. The robustness and accuracy of the models produced by the AMLP were also tested using the testing data sets. The experiments results showed that the proposed scheme for AMLP was able to produce an accurate and robust model increasing the effectiveness of dynamic forecasting of electric load consumption.

REFERENCES

Arroyo, D., M. Skov and Q. Huynh, 2005. "Accurate Electricity Load Forecasting with Artificial Neural Networks," presented at the *2005 International Conference on Computational Intelligence for*

Modelling, Control and Automation, and International Conference of Intelligent Agents, Web Technologies and Internet Commerce.

Badran, S. and O. Abouelatta, 2013. "Forecasting Electrical Load using ANN Combined with Multiple Regression Method," *The Research Bulletin of Jordan ACM.* vol. 2, no. 2, pp. 152-158.

Boniface, Y., F. Alexandre and S. Vialle, 1999. "A Bridge Between Two Paradigms for Parallelism: Neural Networks and General Purpose MIMD Computers," Presented at the *International Joint Conference on Neural Networks*, Washington, D. C., 1999.

Chujai, P., N. Kerdprasop and K. Kerdprasop, 2013. "Time Series Analysis of Household Electric Consumption with ARIMA and ARMA Models," *Proceedings of the International MultiConference of Engineers and Computer Scientists*, Hongkong, vol. 1.

Duda, R. O., Hart, P. E. and Stork, D. G. 2001. "Pattern Classification." New York: John Wiley & Sons, New York.

Edwards, R. E. et al., 2012. "Predicting future hourly residential electrical consumption: A machine learning case study," *Energy Buildings*, 2012. doi: 10.1016/j.enbuild.2012.03.010.

Kandananond, K., 2011. "Forecasting Electricity Demand in Thailand with an Artificial Neural Network Approach," *Energies.* vol. 4, pp. 1246-1257.

Kazemi, A., H. Shakouri, M. B. Menhaj, M. R. Mehregan and N. Neshat, 2011. "A Multi-level Artificial Neural Network for Residential and Commercial Energy Demand Forecast: Iran Case Study," *Proceedings of the 3rd International Conference on Information and Financial Engineering*, Singapore, pp. 25-29.

Lalis, J. T., B. D. Gerardo, and Y. Byun, 2014. "An Adaptive Stopping Criterion for Backpropagation Learning in Feedforward Neural Network," *IJMUE*, vol. 9, no. 8, pp. 149-156.

Lin, F., X. H. Yu, S. Gregor, and R. Irons, 1995. "Time Series Forecasting with Neural Networks," *Complexity International*, Vol. 2, 1995.

Nguyen, D. and B. Widrow, 1990. "Improving the Learning Speed of 2-Layer Neural Networks by Choosing Initial Values of the Adaptive

Weight, " *Proceedings of the International Joint Conference on Neural Networks*, San Diego, CA, USA, vol. 3, pp. 21-26.

Simsar, S., M. Alborzi, J. Nazemi and M. A. Layegh, 2013. "Forecasting Power Demand using Neural Networks Model," *IJEAT*, vol. 2, no. 5, pp. 441-446.

Yang, J., H. Rivard, and R. Zmeureanu, 2005. "Building Energy Prediction with Adaptive Artificial Neural Networks," *Proceedings of the 9th International IBPSA Conference*, Montreal, Canada, pp. 1401-1408.

Yalcintas, M. and S. Akkurt, 2005. "Artificial Neural Networks Applications in Building Energy Predictions and a Case Study for Tropical Climates," *IJER*, pp. 891-901, vol. 29.

ABOUT THE AUTHORS

Elmer Asilo Maravillas, PhD-ME

Affiliation: Full Professor, College of Computer Studies, Cebu Institute of Technology - University Cebu City, Cebu, Philippines

Education: Doctor of Philosophy in Mechanical Engineering (PhD-ME)

Business Address: N. Bacalso Avenue, Cebu City, 6000 Philippines

Research and Professional Experience:
R&D Coordinating Office Consultant

Honors: PhD-ME with Distinctions

Publications from the Last 3 Years:
1. Nogra, James Arnold, Cherry Lyn Sta Romana, Elmer Maravillas (2019). "LSTM Neural Networks for Baybáyin Handwriting Recognition". 2019/2/23, Pp. 62-66. *2019 IEEE 4th International*

Conference on Computer and Communication Systems (ICCCS). Publisher: IEEE.
2. dela Cerna, Monalee A., Elmer A Maravillas (2016). An Application of Partitive Clustering Algorithm for Landslide Hazard Zonation. *Journal Proceedings of the International Multi Conference of Engineers and Computer Scientists*. iaeng.org.
3. Jordan, Chris, G Aliac, Elmer Maravillas.OT Hydroponics Management System. *2018 IEEE 10th International Conference on Humanoid, Nanotechnology, Information Technology, Communication and Control, Environment and Management (HNICEM)*. Pp: 1-5. IEEE.
4. Malangsa, Rhoderick D. and Elmer A. Maravillas. Abaca Tissue Culture Contamination Grading Using Naïve Bayesian Classification. *2017 CEBU International Conference on Advances in Science, Engineering and Technology (ICASET-17)* Jan. 26-27, 2017, Cebu (Philippines). Pp: 7-12. http://doi.org/ 10.17758/URUAE.AE0117211.
5. Oraño, Jonah Flor V., Jomari Joseph A. Barrera, and Elmer A. Maravillas. Jackfruit Phytophthora Palmivora (Butler) Disease Recognizer Using Naïve Bayes Classifier. *Journal of Computational Innovations and Engineering Applications*. July 2018, Pp. 1-7. De La Salle University - Manila.

Jermias T. Lalis

Affiliations:
- College of Computer Studies, Cebu Institute of Technology - University Cebu City, Cebu, Philippines
- Teradata Corp., Bonifacio Global City, Taguig

Education: Doctor in Information Technology

Research and Professional Experience:
- Data Scientist, Teradata Corp., 2017-2019

- Data Scientist, Cash Credit, 2016-2017
- Research Director, La Salle University-Ozamiz, 2015-2016
- Assistant Professor/Researcher, La Salle University-Ozamiz, 2007-2016

Publications from the Last 3 Years:
1. "A Mechanism for Online and Dynamic Forecasting of Monthly Electric Load Consumption Using Parallel Adaptive Multilayer Perceptron (PAMLP)". *International Conference on Computer and Information Science*, 177-188, 2018.
2. "A New Multiclass Classification Method for Objects with Geometric Attributes Using Simple Linear Regression". *IAENG International Journal of Computer Science* 43(2), 198-203, 2016.

In: Multilayer Perceptrons
Editor: Ruth Vang-Mata
ISBN: 978-1-53617-364-2
© 2020 Nova Science Publishers, Inc.

Chapter 5

DEVELOPMENT OF THE PRE-FRACTAL PATCH ANTENNA WITH ARTIFICIAL NEURAL NETWORK

Elder Eldervitch Carneiro de Oliveira[1,*],
Marcelo da Silva Vieira[1],
Rodrigo César Fonseca da Silva[1],
Pedro Carlos de Assis Jr.[1]
and Paulo Fernandes da Silva Junior[2]

[1]Centro de Ciências Exatas e Sociais Aplicadas,
Universidade Estadual da Paraíba,
Patos, Paraiba, PB, Brazil
[3]Programa de Pós-Graduação em Engenharia da Computação
e Sistemas (PECS), Universidade Estadual do Maranhão,
São Luís, Maranhão, Brazil

---

[*] Corresponding Author's Email: elder@ccea.uepb.edu.br.

Abstract

In this paper, a multilayer perceptron artificial neural network with a layer of hidden neurons trained with the resilient backpropagation algorithm, *the network was used to model a Koch pre-fractal patch antenna*. The training set for the electromagnetic characterization of the antenna was obtained through simulations in the ANSYS commercial software by the momentum method. The neural network model proposed in this paper consists of a multilayer perceptron network that is able to predict antenna behavior within a region of interest with low computational cost, with a training of five thousand epoch, and means square error last than 0.0003.

Keywords: patch antenna, Koch pre-fractal, multilayer perceptron, neural network

Introduction

The increase in research in new technologies and modern applications for wireless communications, such as, internet of things, smart grids, biomedicals, aerospatial, Militar and athletics applications require to use antennas with characteristics of: operation for high frequencies, low weight, low cost, directional or omnidirectional antenna-type irradiation, compliance with wavy surfaces and their most diverse applications in microwave devices and circuits, one of the solutions used is the microstrip antennas [1]-[3].

From a mathematical point of view, a fractal refers to a set in Euclidean space with specific properties, such as: self-similarity or self-affinity, simple and recursive definition, fractal dimension, irregular shape, natural appearance [4]. Fractal geometry is the study of sets with these properties, which are too irregular to be described by calculus or traditional Euclidian geometry language [4], [5]. Fractals are resort to conventional classes, such as: geometrical fractals, algebric fractals and stochastic fractals [6]. Two common methods used to generate mathematical fractals are Iterated Function Systems (IFS) and Lindenmayer Systems [4]-[7]. The design of pre-fractals patch antennas has been a subject of great interest to designers and researchers in the field of antennas. Previously published works by the authors have contributed to this research area, showing the miniaturization of inset-feed patch antennas with the use of Koch and Minkowiski pre-fractals [8]-[9]. Pre-fractal patch antennas are defined with two fractal parameters: iteration number (level) and scaling factor; possess a large design region of interest.

One possible solution to the difficulty of finding an analytical model that accurately describes microwave devices such as fractal antennas is the use of artificial neural networks (ANN). According [10] "A neural network is a massively distributed processor made up of simple processing units that have a natural propensity for storing experiential knowledge and making it available for use". In this work is relating level 2 pre-fractal antenna parameters with desired specifications for resonance frequency (Fr), bandwidth (BW), quality factor (Q) and inset-fed length (y0). From the training of the ANN, it is possible to reproduce and/or generalize the knowledge acquired within the region of interest investigated for the proposed antenna, with a low computational cost. A model using ANN with shapes and multiband behavior that facilitate frequency reconfiguration is developed in [9]. The unique properties of geometric fractals are useful to syntesis of more compact patch antenas.

In this paper, the modeling of some electromagnetic parameters of a pre-fractal antenna projected according to the Koch fractal curve in its level 2 triangular geometry proposed in [8] using an artificial neural network. The project uses a multilayer perceptron (MLP) neural network with a layer of hidden neurons and an output neuron layer. The modeling of the electromagnetic parameters was performed in a microstrip patch antenna, for the microwave frequency range. Several works had used ANN for resolution problems in enginering and othes sciencies [11].

This article consists of four more sections in addition to this introduction. Section 2 describes the process of generating the Koch fractal curve as well as the structure of the pre-fractal antenna analyzed. Section 3 briefly describes the neural networks as well as the network architecture used.

Numerical results regarding the modeling of this antenna are presented in Section 4. Section 5 shows the conclusions of this article.

MATERIAL AND METHODS

Fractal Geometry – Koch Curve

The first studies on fractal geometry date back to the last century with the work presented by Polish researcher and mathematician Benoit Mandelbrot [5].

The term fractal figure refers to recursively constructed objects, where a smaller piece of this object is a replica identical to the original. The works cited in [12] and [13] pioneered the application of this geometry in antenna and filter designs for the microwave frequency range. Unlike Euclidean geometry, which has the entire dimension, fractal geometry has a fractional dimension (D), given by:

$$D = \frac{Log(N)}{Log(r)} \tag{1}$$

where N is the total number of distinct copies of the original element and r is the iteration scale factor of the fractal.

Other important features of these fractal figures are:

i) the space-filling property, which is responsible for the process of miniaturization of antennas and other microwave devices;
ii) the self-similarity property that is responsible for the multiband behavior of the antennas.

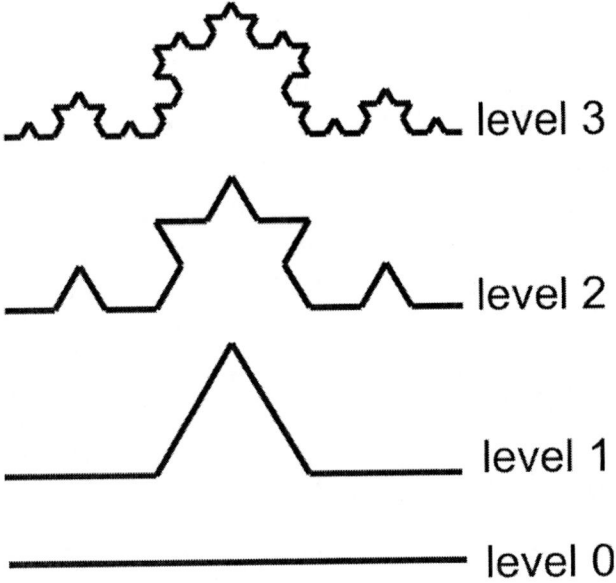

Figure 1. Koch fractal curve for four levels.

As shown in Figure 1, the Koch curve will be obtained through the infinite application of an interactive process applied to an initial figure, in this case, a straight line (called level 0). Initially, this straight line is divided into three equal parts and replaces the intermediate segment with an equilateral triangle without the base, thus obtaining the level 1 Koch curve. The next step is to divide each level of the Koch curve again. 1. replacing the intermediate following by an equilateral triangle without the base, thus obtaining the level 2.

The process is repeated indefinitely to obtain the other levels, at the limit is an ideal fractal.

Typically, for practical projects in the area of microphyte antennas and/or frequency-selective filters, only Koch curve levels 1 and 2 are used because of the difficulty of constructing such devices with higher Koch curve levels.

Artificial Neural Network

An artificial neural network (ANN) can be defined as a machine that is designed to model the way the brain (biological neuron) performs a particular task or function of interest. To achieve good performance, neural networks employ a massive interconnection of processors, also called a neuron [10]. Figure 2 illustrates the model of an artificial neuron. Mathematically a neuron j of an RNA with N_i entries is expressed by:

$$net_j = \sum_{i=1}^{N_i} X_i W_{ji} + b \qquad (2)$$

$$y_i = \phi(net_j), \qquad (3)$$

where $x_1, x_2, ..., x_{ni}$ are the input signals, $W_1, W_2, ..., W_{ji}$ are the synaptic weights between neuron i and neuron j, b is a bias value; net_j is the activation potential; ϕ (.) is the activation function, and y_j is the output signal of the neuron. There are several activation functions that can be employed in formulating a neuron.

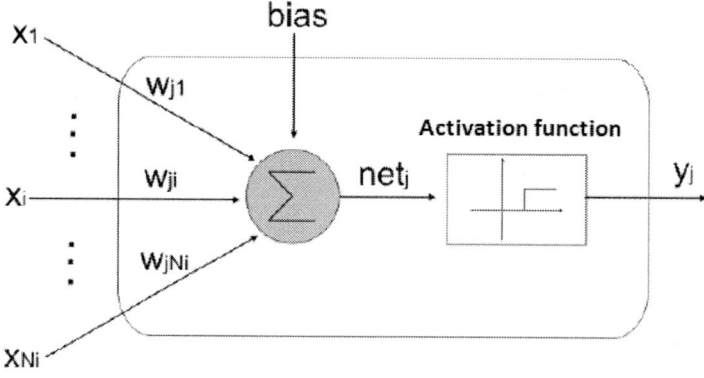

Figure 2. Artificial neuron model.

Generally speaking, the most commonly used activation function is the sigmoid function, given by equation (4):

$$\phi net_j = \frac{1}{1+exp(-net_j)} \qquad (4)$$

An artificial neural network architecture used is the multilayer perceptron, a structure with significant computational capacity, which has in its architecture more than one layer of hidden neurons. This structure is capable of solving nonlinearly separable problems using one or more intermediate layers of hidden neurons [14]. Figure 3 illustrates an MLP network architecture, with an input layer (with three inputs), an intermediate layer of hidden neurons, and an output layer with four neurons. This is the neural network model used in this work for the modeling of the analyzed antenna.

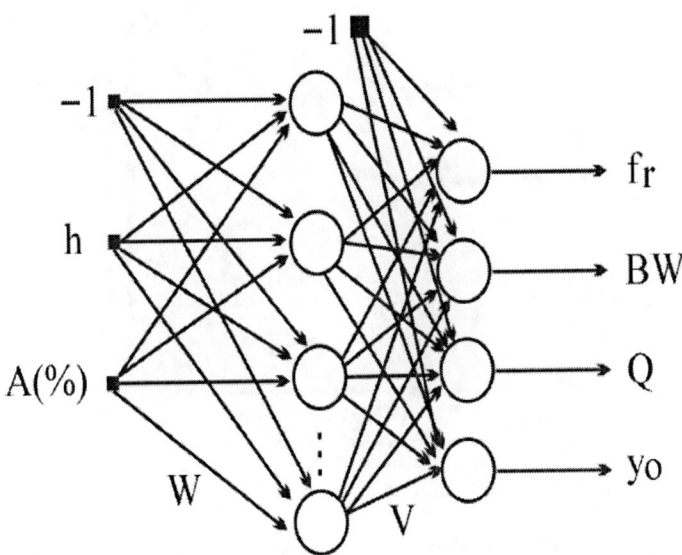

Figure 3. Model of the multilayer perceptron neural network used.

where the inputs are: the bias (-1); the thikness of dieletric (h), A(%) is the antenna dimension related by with of patch element (W). The outputs are, the resonance frequency (f_r), the bandwidth (BW), the factor of quality (Q) and the inset-fed (y_0).

Koch Pre-Fractal Path Antenna

The term pre-fractal is because the modeled structure is not symmetrical in all its dimensions [5]. The geometry of the modeled antenna is show in Figure 4, with its dimensions in millimeters (mm). To the level 2 Koch curve was applied the radiant and non-radiant margins of a conventional rectangular patch antenna with resonance frequency of 2.45 GHz, as shown in [8].

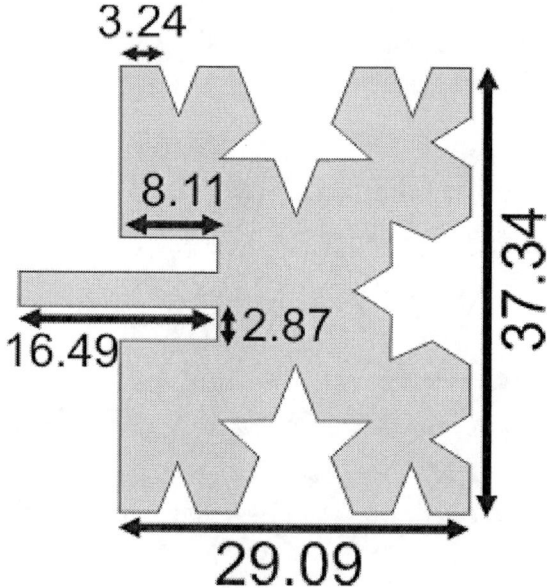

Figure 4. Koch pre-fractal patch antenna modeling.

The antenna being excited consists of a pre-fractal patch element mounted on an isotropic dielectric layer deposited on a ground plane, by a microstrip line connected to a 50 Ω SMA connector. The technique used in this structure, to enhance its impedance characteristics, was inset-fed (y_0).

The substrate used was glass fiber (FR4), thickness (h) of 1.5mm and a relative electrical permittivity (ε_r) of 4.4.

MODELING OF KOCH PRE-FRACTAL PATCH ANTENNA BY ANN

Fractal Geometry – Koch Curve

The neural network model used for modeling the parameters associated with the antenna is the one shown in Figure 3, with three input nodes, with 10 neurons in the hidden layer, and 4 output neurons (corresponding to the modeled antenna parameters). For the modeling of the antenna analyzed, an MLP network was considered which was trained with the resilient backpropagation algorithm [14]. The network training set was obtained by varying the thickness (h) of the substrate, h = [1.5; 3.0; 4.5; 6.0; 7.5; 9.0; 10.5]mm, for each antenna dimension analyzed, A% = [50; 60; 75; 90; 100], consisting of 7 points per curve, totaling 28 samples. In which, the 75% curve was removed for testing and validation of the proposed neural model, whith the simulation results obtained through a parametric study in the ANSYS software.

The electromagnetic data obtained considering the variation of substrate thickness and antenna size, as well as the network response to the imposed training set are shown in Figures 5 – 8. It is observed from Figure 5 an increase of the resonance frequency, with the reduction in antenna size and decrease in the substrate thickness from a value of h = 9.0mm to an antenna size of 50%. It can also be observed that the network learned satisfactorily from the imposed training set, on the interpolating the points of the curve, thus generalizing very well the knowledge acquired within the region of interest.

The result of Figure 6 shows the percentage bandwidth behavior of the antenna with different dimensions by varying the substrate thickness. An exponential behavior is noted in the curve that describes the percentage bandwidth as the substrate thickness increases, with the highest value obtained for the A% = 50 curve with an h = 10.5mm. The network output showed approximate learning and a good ability to generalize the percentage bandwidth to other antenna dimensions within the analyzed region. The quality factor (Q) is a parameter that relates the resonant frequency and bandwidth of an antenna. Figure 7 shows the behavior of this Q factor as we increase the thickness of the substrate used as a function of antenna dimensions. From Figure 7 we can see that the highest Q value obtained was for an antenna with 100% of its physical dimension (A% = 100), with a value of h = 1.5mm. Again, after the training phase, the neural network learned from the imposed training set and was able to interpolate and generalize the knowledge acquired within the range of interest with good precision.

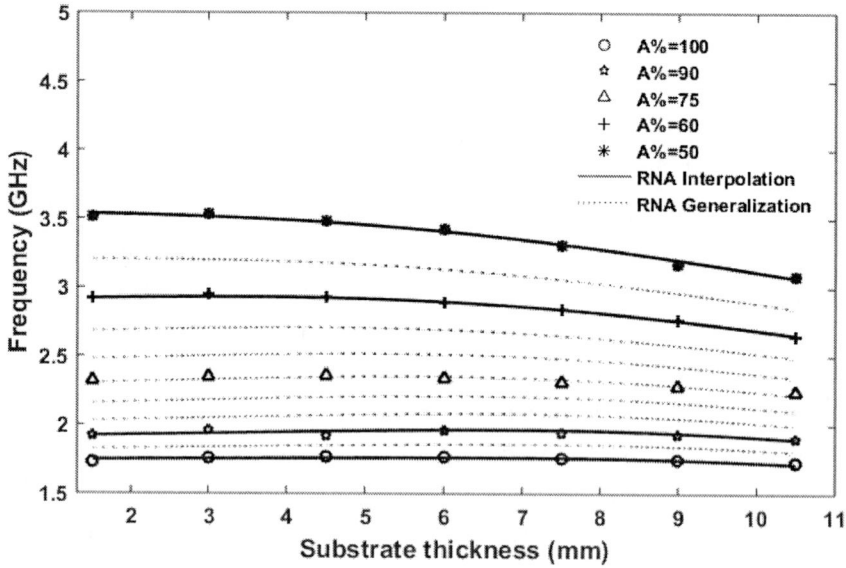

Figure 5. Modeling of the resonance frequency (fr) parameter of the Koch pre-fractal patch antenna.

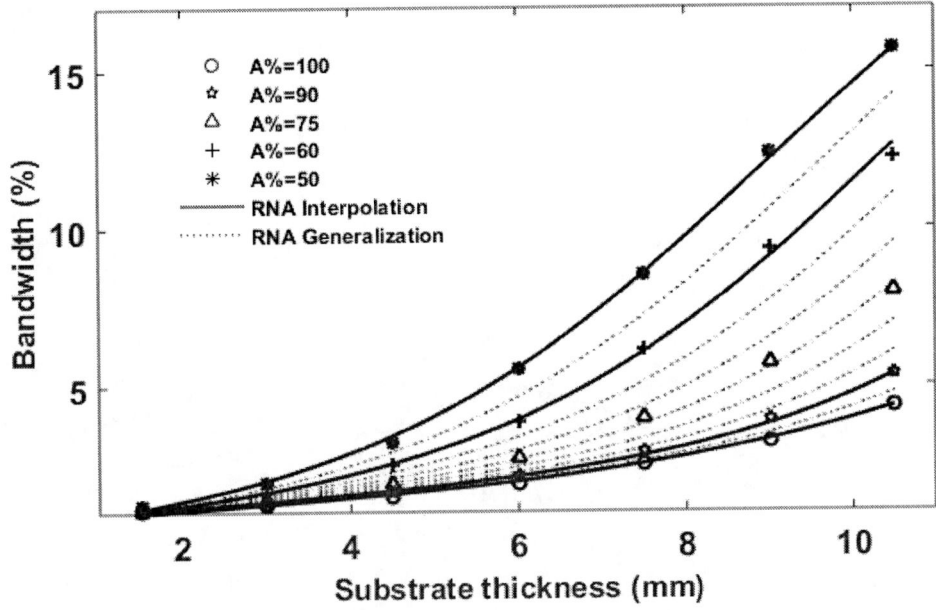

Figure 6. Modeling of percentual bandwidth (BW%) parameter of the Koch pre-fractal patch antenna.

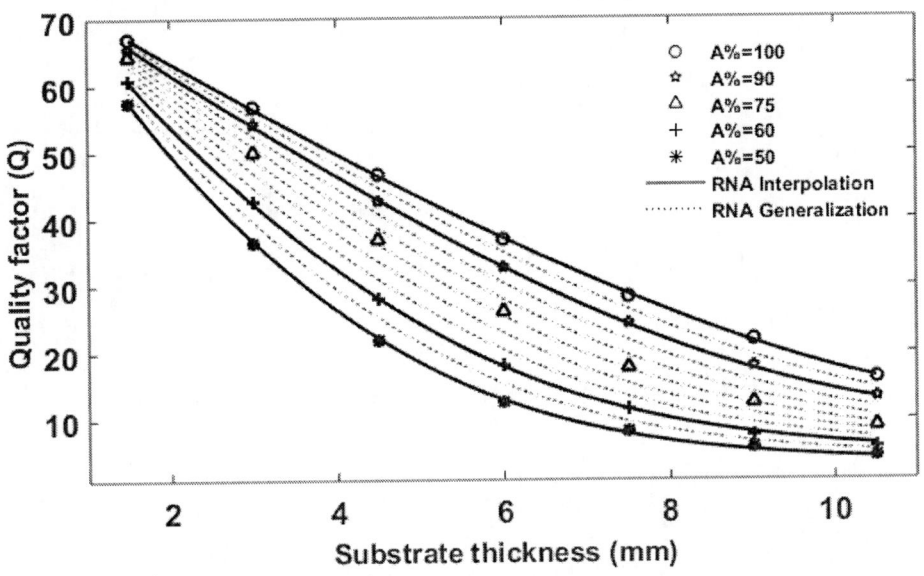

Figure 7. Modeling of quality factor (Q) parameter of the Koch pre-fractal patch antenna.

The result of Figure 8 shows the behavior of the inset-fed parameter (y0) as a function of the variation of substrate thickness used and antenna size. It is possible to observe from the result an increase of the inset-fed length proportional to an increase of the substrate thickness used and the antenna dimension.

The highest inset-fed value observed was for the antenna with dimension A% = 100 and with a value of h = 6.0mm, in then there is a tendency to decrease this value.

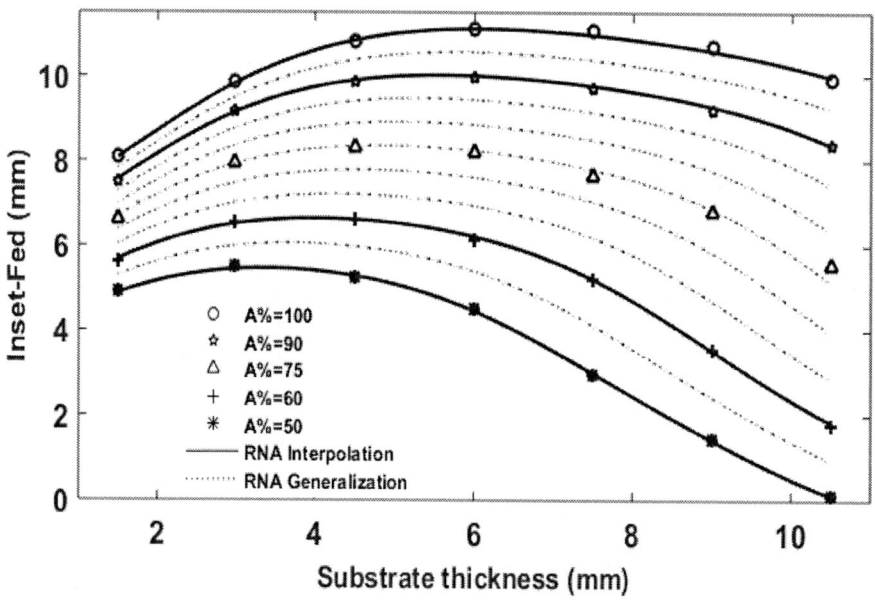

Figure 8. Modeling of the inset-fed (y_0) parameter of of the Koch pre-fractal patch antenna.

Table 1. Simulated results of the mean square error

Modeled Parameters	Training times	Mean square error
Resonance frequency (f_r)	5000	3.63×10^{-5}
Bandwdith (BW)	5000	3.56×10^{-5}
Quality factor (Q)	5000	3.53×10^{-5}
Inset-fed (y_0)	5000	3.82×10^{-5}

We can also observe that the network output for the modeled parameter was satisfactory, in which the network learned from the imposed training set and generalized well within the region of interest to untrained values. Table 1 shows the mean square error obtained after 5000 epoch of the MLP network-training phase proposed for the modeling of the analyzed antenna parameters.

CONCLUSION

In this paper, a multilayer perceptron neural network trained with the resilient backpropagation algorithm which is proposed for modeling the electromagnetic parameters associated with a pre-fractal Koch antenna. For the modeling of the parameters, a training set obtained through a parametric study in the ANSYS software of the proposed antenna was used, in which the substrate thickness and the antenna dimensions were varied. It was observed that the multilayer perceptron network learned from the training set imposed on it, being able to interpolate the points in the training phase and efficiently generalize the knowledge acquired within the analyzed region of interest, proving to be as accurate as of the analytical models with low computational short.

ACKNOWLEDGMENT

The authors of the paper thank CNPq, Paraíba State University (UEPB), the Federal Institute of Paraíba (IFPB), and Maranhão State University (UEMA) for their support in carrying out this work.

REFERENCES

[1] Pandey, Anil. 2019. *Practical Microstrip and Printed Antenna Design.* Norwood: Artech House.

[2] Mitra, Ray. 2018. *Developments in Antenna Analysis and Design,* Vol. 1 and 2. London: SciThech Publising.

[3] Cohen, N. 1997. "Fractal Antenna Applications in Wireless Telecommunications". *Proceedings of Electronics Industries Forum of New England*, pp. 43 - 49.

[4] Sullivan, D. M. 2000. *Electromagnetic Simulation Using the FDTD Method.* New York: IEEE Press.

[5] Mandelbrot, B. B. 1983. *The Fractal Geometry of Nature.* New York: W. H. Freeman and Company.

[6] Falconer, K. 2003. *Fractal geometry: mathematical foundations and application.* 2. ed. Londres: Wiley.

[7] Mishra, J. and S. Mishra. 2007. *L-Systems Fractals.* Amsterdam: Elsevier.

[8] Oliveira, E. E. C., Campos, A. L. P. S. and P. H. F. Silva. 2009. "Quasi-fractal Koch Triangular Antenna". In: *2009 SBMO/IEEE MTT-S International Microwave and Optoelectronics Conference (IMOC).* Belém, November 3 - 6, pp. 163 - 166. doi: 10.1109/IMOC.2009.5427607.

[9] Oliveira, E. E. C., Silva, P. H. F., Campos, A. L. P. S. and S. G. Silva. 2009. "Overall size antenna reduction using fractal elements". *Microwave and Opt. Technol. Letters,* doi: 10.1002/mop.

[10] Haykin, Simon. 2009. *Neural Networks and Leaning Machines.* New Jersey: Person.

[11] Cain, Gayle. 2017. *Artificial Neural Networks: New Research.* New York: Nova Publishers.

[12] Puente, C., Romeu, J., Pous R. and A. Cardama. 1998 "On the Behavior of the Sierpinsk Multiband Fractal Antenna". *IEEE Transactions on Antennas and Propagation,* vol. 46, pp. 517 - 524.

[13] Mutiara, A. B., Reflanti, R. and Rachmansyah. 2011. "Design of Microstrip Antenna For Wireless Communication at 2.4GHz". *Journal of Theoretical and Applied Information Technology*, vol. 33, pp. 184 - 192.

[14] Ridmiller, M. and H. Braun. 1993. "A direct adaptive method for faster backpropagation learning: the RPROP algorithm". *Proceedings of the IEEE International Conference on Neural Networks*, San Francisco, pp. 586 - 591.

INDEX

A

accelerometer, 40, 47, 48, 53, 56, 57, 58, 60, 66, 67, 68, 69, 70, 71, 72, 74, 75
adaptive multilayer perceptron, v, vii, ix, 101, 102, 104, 109, 110, 119, 123
age estimation, v, ix, 77, 78, 79, 96, 97, 98, 99
algorithm, vii, viii, x, 6, 7, 10, 12, 15, 17, 18, 20, 21, 22, 26, 27, 29, 35, 37, 41, 43, 44, 45, 46, 56, 58, 60, 71, 78, 85, 86, 88, 100, 105, 108, 126, 133, 137, 139
Artificial Neural Networks, 100, 119, 121, 138

B

backpropagation, vii, viii, x, 43, 44, 78, 100, 102, 105, 120, 126, 133, 137, 139
backpropagation algorithm, vii, viii, x, 44, 78, 105, 126, 133, 137
bandwidth, 127, 132, 134, 135
bias, 11, 21, 22, 27, 32, 130, 131
boundary value problem, viii, 1, 2, 3, 8, 10, 17, 18, 27, 35, 36

C

Cloud computing, 40, 45, 46, 50, 56, 60, 61
cloud computing environment, 40

D

deep brain stimulation, 40, 41, 53, 54, 56, 57, 58, 59, 60, 61, 62, 63, 64, 66, 73, 74, 75
differential equations, vii, viii, 1, 2, 3, 10, 11, 12, 21, 26, 27, 32, 33, 34, 35, 36, 37

E

economic development, 102
economic indicator, 104
electric load consumption, v, ix, 101, 102, 104, 119, 123
electricity, 102, 104
electromagnetic, x, 126, 127, 133, 137
energy, ix, 101, 102, 103, 104, 118
energy consumption, 103, 105
essential tremor, 54, 63, 64, 72, 73, 75

estimation process, ix, 77, 96
Euclidean space, 126
Euclidian geometry, 126
evolution, vii, viii, 12, 40, 41, 47, 53, 54, 60

F

finite element method, 26
forecasting, vii, ix, 101, 102, 103, 104, 105, 106, 109, 119
fractal dimension, 126

G

gait, 40, 41, 50, 51, 61, 62, 63, 64, 67, 71, 72
gyroscope, 40, 48, 50, 51, 60, 70, 72, 75

H

hemiparesis, 50, 51
Human Development Index, 102

I

image processing techniques, v, 77, 78, 99, 100
initial value problem, viii, 1, 2, 6, 7, 12, 17, 37
Internet of Things, viii, 39, 40, 60, 62

K

Koch pre-fractal, viii, x, 126, 132, 133, 134, 135, 136

L

learning, viii, 7, 9, 11, 39, 40, 41, 42, 45, 46, 47, 48, 49, 50, 51, 52, 54, 56, 58, 59, 60, 61, 64, 66, 70, 72, 75, 86, 88, 89, 92, 103, 105, 108, 109, 120, 134, 139
learning process, 11, 48
long-term forecasting, 102, 103, 104

M

machine learning, v, viii, 39, 40, 41, 42, 45, 46, 47, 48, 49, 50, 51, 52, 54, 56, 58, 59, 60, 61, 64, 65, 66, 70, 72, 75, 76, 86, 120
mathematical fractals, 126
models, vii, ix, 3, 37, 101, 103, 106, 108, 109, 113, 114, 119, 137
movement disorder, 40, 41, 53, 54, 60, 61, 62, 63, 64, 66, 72, 73, 74
multilayer perceptron, v, vii, viii, x, 1, 3, 5, 7, 9, 11, 13, 15, 16, 17, 19, 21, 22, 23, 25, 27, 29, 31, 33, 35, 37, 39, 40, 41, 42, 43, 45, 46, 47, 48, 49, 50, 51, 52, 53, 54, 56, 58, 59, 60, 61, 70, 72, 75, 78, 100, 104, 105, 109, 126, 127, 131, 137
multilayer perceptron artificial neural network, v, vii, viii, x, 1, 3, 5, 7, 9, 11, 13, 15, 17, 19, 21, 23, 25, 27, 29, 31, 33, 35, 37, 126
multilayer perceptron model, 78
multilayer perceptron neural network, v, vii, viii, 23, 35, 39, 40, 41, 42, 43, 45, 46, 47, 48, 49, 50, 51, 52, 53, 54, 56, 58, 59, 60, 61, 70, 72, 75, 100, 131, 137

N

Network Centric Therapy, v, viii, 39, 40, 41, 45, 46, 47, 50, 51, 53, 54, 55, 60, 61, 62, 63, 64 neural network, v, vi, vii, viii, x, 1, 2, 3, 5, 9, 10, 11, 12, 13, 15, 16, 17, 18, 21, 22, 23, 24, 25, 26, 27, 28, 29, 31, 32, 33, 34, 35, 36, 37, 38, 39, 40, 41, 42, 43, 44, 45, 46, 48, 49, 50, 51, 52, 54, 56, 58, 59, 60, 61, 62, 70, 72, 74, 75, 77, 79, 85, 86, 87, 88, 91, 92, 96, 99, 100, 103, 104, 105, 106, 107, 119, 120, 121, 125, 126, 127, 130, 131, 133, 134, 137, 138, 139

O

ordinary differential equations, 1, 2, 3, 10, 34, 36, 37

P

panoramic graph, 78
partial differential equations, vii, viii, 1, 3, 26, 32, 33, 34, 35, 37
patch antenna, vi, vii, viii, x, 125, 126, 127, 132, 133, 134, 135, 136
patellar tendon reflex, 40, 47, 48, 49, 67, 68, 70

Q

quantification, 48, 53, 60, 63, 66, 68, 69, 70

R

reflex response, 40, 41, 47, 48, 50, 60, 62, 63, 64, 69, 70

S

sensor, 40, 41, 46, 47, 53, 54, 61, 62, 66, 75

software, x, 41, 48, 51, 56, 58, 60, 61, 65, 126, 133, 137
stimulation, 40, 41, 48, 54, 56, 57, 58, 59, 60, 61, 64, 66, 74, 75
stochastic fractals, 126
sustainable development, 103

T

techniques, vii, ix, 3, 11, 16, 26, 28, 37, 78, 79, 96, 100
technology(ies), 74, 126
technology, 74
testing, ix, 41, 54, 73, 80, 90, 96, 101, 112, 116, 119, 133
time series, vii, ix, 101, 106, 109, 110, 111, 112
transmission, 42, 45, 56

V

valuation, 50, 62, 63, 64, 66
vector, 22, 26, 29, 31, 48, 54, 79, 84
velocity, 19, 29

W

Waikato Environment for Knowledge Analysis (WEKA), 41, 42, 43, 48, 49, 51, 52, 56, 57, 58, 59, 60, 65
wearable inertial sensors, 40
wireless connectivity, 41, 55
wireless inertial sensors, 40
wireless systems, viii, 39, 40, 41, 50, 60, 61